THE CORNCRIBS OF BUZET

THE CORNCRIBS OF BUZET

MODERNIZING AGRICULTURE IN THE FRENCH SOUTHWEST

Peter H. Amann

PRINCETON UNIVERSITY PRESS   PRINCETON, NEW JERSEY

Copyright ©1990 by Princeton University Press
Published by Princeton University Press, 41 William Street,
Princeton, New Jersey 08540
In the United Kingdom: Princeton University Press, Oxford
All Rights Reserved

Library of Congress Cataloging-in-Publication Data

Amann, Peter H., 1927–
The corncribs of Buzet : modernizing agriculture in the French
southwest / Peter H. Amann.
p.   cm.
Includes bibliographical references.
ISBN 0-691-05563-7 (alk. paper)
1. Agriculture—Economic aspects—France—Buzet-sur-Tarn Region.
2. Agricultural innovations—France—Buzet-sur-Tarn Region.
I. Title.
HD9012.8.B89A42   1990
338.1'0944'85—dc20   89-70116   CIP

This book has been composed in Linotron Caledonia

Princeton University Press books are printed
on acid-free paper, and meet the guidelines
for permanence and durability of the Committee
on Production Guidelines for Book Longevity
of the Council on Library Resources

Printed in the United States of America by
Princeton University Press, Princeton, New Jersey

10 9 8 7 6 5 4 3 2 1

To Enne

Contents

Illustrations

Abbreviations Used in Text

AOC	Appellation d'origine contrôlée
CETA	Centres d'Etudes Techniques Agricoles
CUMA	Coopérative d'utilisation de matériel agricole
DDA	Direction départementale de l'Agriculture
EEC	European Economic Community
FEOGA	Fond européen d'orientation et de garantie agricole
GAEC	Groupement agricole d'économie en commun
INRESA	Institut de recherche scientifique agricole
IVD	Indemnité viagère de départ
SAFER	Société d'aménagement foncier et d'établissement rural
SMIC	Salaire minimum interprofessionel de croissance
VDQS	Vin délimité de qualité supérieure

Preface

AFTER two years of research in southwestern France in 1981–1983, a further couple of years devoted, on and off, to transcribing interviews and three years of intermittent writing, it is delightful to have reached the point where I can acknowledge the help of strangers, family, friends, and colleagues who, by providing information or criticism or both, contributed to the manuscript on which this book is based. My first and most obvious debt is to the farmers, farm wives, technicians, and mayors of Haute-Garonne who hospitably received an inquisitive foreigner (only once was I suspected of being a CIA agent!) and talked to him with considerable candor. I owe a special debt to M. Jean-Claude Germain, who doubled as both informant and, thanks to his knowledge of English, critic, in connection with my chapter on Saint Cézert. It is also a pleasure to thank my wife, Enne, for her repeated critical readings, our son David, as well as Professors Melvin Cherno, Père Olivier de Framond, S.J., Jean Peneff, Don Proctor, Richard Roehl, and Sid Warschausky for their critical efforts. Despite the aid I received, I take full responsibility for the final content of this study and for whatever errors of fact and interpretation I may have overlooked.

Like many scholars I am also indebted both to my university and to foundations that have generously supported this research. My initial year in Toulouse was funded by a sabbatical from the University of Michigan-Dearborn and I also appreciated a later minisabbatical that gave me a clear stretch of time permitting me to complete two-thirds of this book without interruption. I received a fellowship for 1982 from the National Endowment for the Humanities that permitted me to extend my research stay in France for a second year. I was also awarded a grant-in-aid from the American Philosophical Society that defrayed the considerable costs of copying records and documents that I would not have had the time to analyze otherwise. In a somewhat different context, I also wish to acknowledge the permission granted to me by the Institut géographique national (IGN), France, to reproduce and enlarge topical maps of the five *communes* with which this book deals.

It is equally traditional to acknowledge the help of professionals and institutions without whose cooperation historians are virtually helpless. M. Ropars, the now retired *documentaliste* of the Chambre d'Agriculture of Haute-Garonne, went out of his way to be helpful, as did M. Pradine of the Société d'aménagement foncier et d'établissement rural (SAFER).

Agricultural extension agents, active and retired, all took an interest in my research, but M. Roblin of Blagnac and M. Garrigues of Villemur-sur-Tarn outdid themselves in laying out the problems and prospects of agriculture in their respective bailiwicks. The staff of the Direction départementale de l'agriculture of Haute-Garonne and the region Midi-Pyrenées threw open its archives, made endless copies for me and gave me all the guidance I needed, even though these were records normally closed to scholars. The same may be said for the officials of the Bureau du remembrement of the Ministry of Agriculture in Paris, even though, because of the shifting focus of my study, I did not have occasion to incorporate what I found there into this book. I would have been severely handicapped had I not been able to utilize the holdings of the Toulouse Municipal Library, as well as those of the University of Toulouse library. My much briefer consultation of the departmental archives of Gers and of Tarn (both in connection with what turned out to be tangential questions concerning Cadours), as well as of the National Archives in Paris, was facilitated by the personal interest of archivists in the provinces and by the usual professionalism in Paris.

Obviously, I could never have been able to complete this study without access to the departmental archives of Haute-Garonne. For all but the *commune* of Loubens-Lauragais, I drew the bulk of my information on social and agricultural conditions prior to the last two generations from these archives. At the same time, no doubt like other scholars in similar circumstances, I often felt stymied because large portions of their holdings are uninventoried. Just as frequently I was frustrated because so many documents that, logically, should have been available, had been discarded somewhere along the line or else been lost in a conflagration during World War II. Still, whatever we as interested foreign observers or resident aliens may think of *les lourdeurs de l'administration française*, as historians we can only be grateful that the weight of this bureaucracy has left such an extraordinarily full record of the past that even its scattered fragments keep us going professionally.

Peter H. Amann
Ann Arbor, March 1988

THE CORNCRIBS OF BUZET

Introduction _____

It is hardly a secret that everywhere in the Western world, the pace of socioeconomic change in rural communities has accelerated fantastically over the last few decades. This is true throughout western and southern Europe, but nowhere more sharply than in France. Everywhere there has been a decline in the farm population, a consolidation of smaller into larger farms, the final triumph of market over family subsistence agriculture, the substitution of motorized for animal traction, the introduction of new generations of implements, machinery, fertilizers, fungicides, and herbicides, in short, the utilization of a combination of new farming techniques that has led to an unprecedented intensification of production, an immense outpouring of farm products. Putting it abstractly, a low-investment, low-yield agriculture has been transformed into one that combines high investment and high yield. For France, while historians may still argue about just when peasants turned into Frenchmen, no one contests that most French peasants have ended up as market-oriented farmers only in the years since 1945.[1] The somewhat fuzzy term "modernization" will have to do as a convenient shorthand expression summing up this process.

The broad outlines of this vast transformation are well known. French scholars, often working in teams, have since the 1970s excelled at producing historical compendiums of their rural and agricultural history that pay special attention to the revolutionary changes since World War II.[2] From the 1950s on, France has also been home to a school of rural sociologists, among whom Professor Mendras is the best known, committed to analyzing the societal consequences of the current agricultural revolution.[3] Although this book owes a considerable debt to both groups of scholars, it does not fit comfortably into either scholarly tradition.

To the distress of my more social science-minded colleagues, I did not begin my quest with a hypothesis to verify or invalidate, but with open-ended curiosity. I started, of course, with an appreciation that we were witnessing a spectacular socioeconomic revolution in French agriculture, the scope and direction of which were clear enough, even if its destination was not. What aroused my curiosity was how such a sea change was personally experienced by those who lived through it. Precedents from earlier social transformations were not encouraging: for instance, much as we have learned about alterations in the objective conditions of life brought on by the industrial revolution of the eighteenth and nineteenth centuries, we can still only surmise—vide E. P. Thompson's monumental, but necessarily speculative, *The Making of the English Working Class*—how this process was perceived by those who were swept up in it. This is because historians have too often delayed studying social movements until their participants are dead. And unfortunately there are many questions to which documents on which historians traditionally rely do not speak at all.

My chief ambition has been to examine the historical experience with agricultural modernization of French local rural communities and of the farmers who make up such communities and to do so as perceptively, yet also as directly, as possible. As a historian, I had every intention of evoking the historical context of these communities by traditional archival research, but I was even more interested in listening to what farmers had to say about the social transformation that they had experienced in their own lifetimes. As a sympathetic, yet objective, observer, I saw my task as putting their experience into a broader context, of generalizing without reducing what I had been told to colorless anonymity, of reconstructing collective from individual experience without homogenizing that experience. My subsidiary ambition was to convey some of the feel of what I consider to be a dramatic chapter in French, and European, social history. Even aside from the fact that French agricultural statistics generally should be classified with fiction, I did not want to end up with arid commentary on computer-generated graphs and quantitative tables. To those who live through a massive transformation, social change is a vivid, sometimes exhilarating, often frightening, and always gut-wrenching experience. I see no good reason why historians should serve it up sanitized and dehydrated or why it must be addressed to an audience exclusively comprised of specialists. With general readers in mind, I have therefore curbed my Francophile impulses and avoided untranslated French in the text of this book. Similarly, it is I who have perversely insisted on having the scholarly apparatus in the back, so as not to clutter up the story that I had to tell. In connection with my scholarly notes, information derived from interviews has not been documented in that I had promised anonymity to my informants, whose real names are listed in the first note of each chapter, but who are introduced under fake names in the narrative.

Early on, I decided that my approach would be comparative, that I would write parallel studies of distinct rural communities and weigh their similarities and differences. I wanted to draw my conclusions from the empirical evidence, rather than merely verify or refute hypotheses that I had brought to the study. For reasons of practical convenience, I decided to deal with villages in a single French *département*.

My motives for choosing that of Haute-Garonne were various. I wanted to study an area where the most dramatic changes had occurred

1. The *département* of Haute-Garonne within France

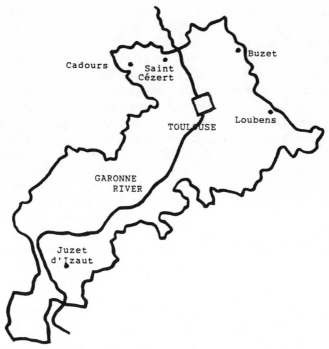

Cadours

Buzet

Saint
Cézert

Loubens

TOULOUSE

GARONNE
RIVER

Juzet
d'Izaut

2. The five townships within the *département* of Haute-Garonne

within the lifetimes of farmers still active, a choice that excluded much
of northern France. As I was to discover in Haute-Garonne, most middle-aged farmers who spoke to me, many of whom were using state-of-
the-art equipment and methods, had begun their farming careers plowing behind a yoke of oxen, occasionally still using the old Roman plow. I
wanted a region with solid agricultural potential, but one neither especially progressive, nor notoriously backward. I also sought a *département*
that had considerable variety as to soil, topography, and types of agriculture. Haute-Garonne ranges from the mountain pastures of the Pyrenees
to the grain basket of the undulating Lauragais plain, its soil from rich,
water-retaining loam to gravel, its farming from stock raising and dairying
through viticulture and polyculture to irrigated corn, sunflowers, and
soybeans. Finally, I thought that it would be helpful if I chose a section
of France where I knew my way around geographically and linguistically.
In retrospect, I continue to think that the *département* of Haute-Garonne
was a good choice, though other *départements* would no doubt have done
just as well.

When I first conceived this project, my focus was somewhat narrower
than it became later: under the impression that "modernization" was too
vague a notion to pursue directly, I thought that concentrating on one
particularly crucial aspect would be more productive. Specifically, I had
decided to focus on government-sponsored local land consolidation—
what the French call *remembrement*—intended to remedy the inconvenience of dispersed fields. Such land consolidation, I was convinced, was
the key to agricultural modernization, because it radically restructured
landed property to make modern farming methods possible.

I soon learned that I had been mistaken, that even though land consolidation did play a significant role in transforming agriculture, it was
not as decisive a turning point as I had assumed. I also discovered that in
the very concrete terms in which farmers conveyed their experience to
me, agricultural modernization was not nearly so nebulous a concept as I
had feared. I concluded that my study would gain from broadening my
perspective to comprise every aspect of the great change that had altered
my French farmers' lives.

In selecting specific communities to study, I took a number of different
factors into consideration. Since my selection took place while I was still
largely concerned with the local response to land consolidation, I chose
only communities in which *remembrement* had been undertaken or attempted. Yet by 1982 only 20 percent of the villages of Haute-Garonne
had participated in that process. It will therefore readily be assumed that
the communities from which I made my selection were automatically
atypical in being more "progressive" than the regional norm. This, however, turns out to be a questionable assumption, because *remembrement*

is not always or even usually synonymous with exceptionally progressive farmers. In some instances, for example, *remembrement*, by definition a political enterprise, was initiated by the mayor, who was rarely personally active in agriculture, rather than by local farmers. There are also examples of communities, where the most progressive agriculturists, who have already consolidated their farms by their own efforts, will oppose *remembrement* as superfluous. In other cases, townships asked to be consolidated because nearby villages, which may have been "progressive," had pioneered earlier, until in a given region *remembrement* came to be taken for granted, as was the case of the Lauragais. Furthermore, there is no necessary correlation between progressive-minded farmers and state-sponsored land consolidation because there are other, more informal, alternatives available for achieving the same ends. Still, had I to do it all over again, I would not confine myself to *communes* having undergone authoritative land consolidation.

Aside from drawing "my" townships from the pool of those officially consolidated, my major concern was to sample different types of agriculture and agricultural regions. One representative community from the Lauragais, a region noted for the best grain-growing soil in southern France, was a must. Since almost the entire acreage of the Lauragais has, as I noted, been consolidated, I could have selected any one of several dozen villages. In the case of Loubens-Lauragais, I was swayed by the availability of notarial records, a very competently drawn up village monograph (there are such monographs, sponsored by the Ministry of Education in the mid-1880s, for virtually all French communities, but their quality varies widely), and, I must confess, by the unique architectural charm of its half timbered houses and its picturesquely dingy *château féodal*. I sought one village from the northeastern quadrant of the *département*, roughly corresponding to the valley of the Tarn River. I picked Buzet-sur-Tarn quite arbitrarily, though I knew that its location within easy commuting distance of Toulouse would pose intriguing problems of suburbanization not found in more remote communities. What I did not know in advance was that among those in the know, Buzet enjoyed a reputation as one of the most advanced and enterprising farming communities in the French southwest. Of mountain villages, only two had actually undergone *remembrement*. Since Soueich was located merely in the foothills of the Pyrenees, while Juzet d'Izaut was a genuine example of mountain agriculture, I had no hesitation in choosing the latter. There were a number of communities in the Gascon hills that qualified. I opted for Cadours when I learned that its former mayor had published a personal account of land consolidation in his village. Looking for a township of winegrowers, I chose Saint Cézert over the better-known Villaudric or Fronton, because in the latter two communities, *remembre-*

ment had been terminated not long before I began my research. Until I actually arrived at Saint Cézert, I had no inkling that, ironically, most of its vintners had meanwhile decided to pull up their grapevines and to convert to irrigated corn and sunflowers. I also completed research on the township of Carbonne as a representative *commune* in the valley of the Garonne, but, in the face of what I sensed as diminishing returns, I decided to drop this sixth community from inclusion in this study. After some hesitation, I decided not to disguise the names and locations of the villages in this study to avoid introducing an unnecessary element of fakery into the proceedings.

Although the use and abuse of oral history has by now become accepted, we deem it still sufficiently novel (a misconception that must make poor Thucydides wince in Hades) to warrant a few comments. About two-thirds of the book deals with socioeconomic change at the grass roots—for once the cliché does apply—since the end of World War II. Even for this contemporary period there are useful printed and manuscript sources that I have utilized: the results of at least three agricultural surveys (1954–1955, 1970, and 1979); yearly lists of active and retired farmers kept by the regional Chamber of Agriculture; occasional working papers, surveys, or monographs on a subregion written up by agricultural extension agents; census lists that are nominally, but only nominally, out of bounds for researchers; voluminous files on every aspect of local land consolidation kept by the regional office of the Ministry of Agriculture; municipally preserved land registers and evidence of land transfers; the mandatory local yearly tenancy reports, and so forth.

Having made my obligatory genuflection in the direction of this Mecca of archival research, it is only fair to add that the major contribution that this book may make to our understanding of French agricultural modernization derives from open-ended interviews with active and retired farmers, agricultural technicians, mayors, and other participant-observers and not from the written record. Altogether I interviewed about one hundred people, the interviews ranging from forty minutes to well over three hours, with something like an hour and twenty minutes as the average. In each locality I began by introducing myself and my project to the village's mayor (not one of whom was a full-time, active farmer, though two were ex-farmers and one farmed part time) and enlisting his cooperation. In the cases of *communes* where the number of active farmers was well under twenty (Saint Cézert, Juzet d'Izaut, and Loubens-Lauragais), I made every effort to contact every working and some retired farmers, as well as the mayor and the cantonal agricultural extension agent. In the case of larger villages (with thirty-five to forty-five active farmers, as in Buzet-sur-Tarn, Cadours, and Carbonne) in consultation with both the local agricultural technician and the mayor, I constructed a sample of about fifteen to eighteen farmers, representing a fair cross-section of the local farming community. Typically, about one farmer in every ten or twelve refused to be interviewed. I myself excluded individuals who had been described to me as *des cas sociaux* for being alcoholic, retarded, or psychotic.

To what extent could I trust the information developed by my informants? Early on, I discovered that if I began an interview in the bureaucratic manner recommended by manuals of oral history, namely with the vital statistics of the respondent, this approach colored all that would follow: my battery of questions would predictably trigger laconic and uninformative answers. As full-fledged members of a bureaucratic society, the French know how to play the bureaucratic game by rote, and the first rule of the game is, "don't volunteer information." Act like a bureaucrat, it became obvious, and you'll get the kind of answers that bureaucrats are thought to deserve.

From the start, I also realized that interviews could be torpedoed by violating my respondents' sense of personal privacy. I became aware that the people with whom I dealt made a sharp distinction between what they saw as their public sphere, their role as farmers and the part they played in their community, and their private universe centered on the family. In their eyes, my expressed interest in how they had coped with what all of them realized was an agricultural revolution was legitimate; similar interest in how this revolution had impinged on their private and family lives would have been resented as intrusive. I therefore systematically restrained myself from raising some issues that did interest me. If my informants in the course of their conversation overcame their reticence, as they sometimes did, so much the better, but any breach of decorum on my part would, for all practical purposes, have terminated the interview.

If on the other hand, I ignored the bureaucratic niceties and respected my informants' sense of privacy, I could place a good deal of reliance on southern French sociability. Once people had agreed to be interviewed, I could count on the norms of a society that greatly prizes articulate self-expression. In practice, this meant letting farmers talk about their agricultural careers as freely and fully as possible, only interrupting with questions when the conversation flagged or when aspects of agricultural development that interested me were touched on too gingerly. Obviously, the more I had already learned from other interviews, the more searching I could make my own questions, which in turn tended to "loosen up" the individual(s) I was interviewing. I became habituated to being tested, to being tried out for size, since it was one thing to pawn off conventional verities on an uninformed outsider; it was more difficult and less gratifying to fudge with an interviewer who was seen as having other sources of information by which what was being said could be checked.

There are, I found out, differences between dealing with written documents and with information developed orally. Historical criticism of written—a synonym for fixed, set, frozen—sources is part of our normal training as historians. It does take a while to learn that oral sources must be treated differently, that they are not unchanging, that, indeed, they may be peeled like an onion. Many good interviews had just this onion-like quality, as different layers of information were revealed as my respondents were carried away by their own story telling and/or reacted to their gradual realization that I was better informed then they had assumed. Sometimes family dynamics, really family dialogue—husband/wife, brother/brother, brother/sister, younger and older generation—complementing and correcting individual statements, uncovered new layers of recollection. Ultimately, to be sure, information derived from interviews must be subjected to the same sorts of tests as any other source, but the dynamics of interviewing themselves have no counterpart in the conventional criticism of written sources.

If we discount differences in intelligence, memory, personality, perceptiveness, articulation, integrity, and the ability to observe, I could count on being able to reconstruct the experience of individuals and their community in considerable depth, particularly when it came to the recollection of individual farming careers. My subjects enjoyed reminiscing for someone obviously interested in what they had seen and learned in their lifetime. The fact that I was a total outsider, a parachutist, so to speak, probably made me less of an object of suspicion than had I been French and thereby occupied a clearly defined place in the social, political, and geographical ranking order of the French. The spoken word, I have come to believe, conveys what ordinary people feel and how they see their world with a sharpness and a density matched among written sources only by the most outspoken letters and the most candid diaries.

This not to say that all aspects of oral recollections are equally trustworthy. If twenty-five years ago, a French North African newcomer to a village of the French southwest was struck by its primitive agriculture, in recalling his shock he may remember how he was the one to introduce the first tractor—when a contemporary farm survey shows conclusively that by that date a number of tractors already were in use in that very community. Memory has a trick of converting generalizations into appropriate, but not necessarily accurate, detail. Conversely, oral recollections of specific dates are notoriously unreliable. Only if, for example, the purchase of the first tractor can be linked to a memorable, dated event—the great freeze of 1956, the return from military service—can most respondents attach a precise date to what everyone recognizes was a major turning point. I also had to make good on my promise to keep the information I received from my informants in confidence: the names mentioned in the chapter are fictitious, though they refer to authentic individuals. In short, oral sources have limitations that must be respected.

My study does have one serious, if unavoidable, shortcoming. One central aspect of the intense agricultural modernization in France over the last three or four decades has been a quasi-Darwinian struggle for survival among agriculturists, which, in most of the villages that I studied, has reduced the farm population to about one-third of what it was in 1945. The mechanism of selection was not so much that unsuccessful peasants left the land, but that, looking like prospective losers, they could not find a wife, or if they had managed to beat these particular odds, that they could not convince their offspring to take over a "hopeless" farm. Consequently, most of the losers in these agricultural wars of attrition have by now been left on the battlefield, while their heirs, if any, have settled into urban jobs in Toulouse. Neither the dead nor the absent can be readily interviewed. This does mean that, like it or not, my study is skewed toward the winners, or at least toward the survivors. To reflect postwar reality faithfully, I would have had to talk to at least two losers for every winner.

I was conscious of the problem. Indeed, I did all I could to seek out the handful of remaining farmers who were hanging on by their teeth and whose farms were unlikely to survive their present operators' retirement. Yet I could not get away from the fact that in 1982 and 1983 when my interviews were conducted, most of the farmers still around—the *commune* of Juzet d'Izaut was the one striking exception—were those who had made it. The majority of those I interviewed were therefore among the Darwinian "fittest," a majority now, but a minority of those who had started out in 1945. For the most part, only the winners were left to tell their tale. Because I had no choice but to adopt their perspective, I have probably inadvertently filtered out a good deal of the loneliness, misery, and disappointment that agricultural modernization brought to the majority of peasants who were unable to make the transition to modern farming.

One final comment on my approach is necessary. It is impossible to talk about agricultural modernization and the experience of farmers without talking a fair amount about farming as such. My own practical experience of farming was confined to one endless summer, when, at the age of fifteen, I worked as a hired man on a Pennsylvania dairy farm, earning $6.00 a week, plus room and board. The only thing I consciously retained from that none too sunny experience was the conviction never to turn my back on a bull calf. Nor can I claim any academic grounding in agronomy to make good my practical deficiencies. Happily, my ignorance did not

prove to be a permanent handicap in France. Every interview provided
practical instruction in farming, at least in farming as practiced in Haute-
Garonne, and everyone was delighted when I evinced interest in the fine
points of their occupation. Given my own informal way of learning something about farming, I have no inclination to stuff prospective readers
with technical details. I have brought such details into this story only
insofar as they are essential to grasping just what agricultural modernization has meant.

In dealing with measurements, I have adopted a not very systematic
compromise between strict adherence to the metric system and the use
of our own American medieval nonsystem. I have been rather casual in
using either "kilometers" (.62 miles) or "miles" for long distances, "meters" (about 1.1 yards or 3.3 feet) or "feet" for short distances. I occasionally have used "acres" when it is not intended to be a precise measure,
but have otherwise stuck with "hectares" (100 meters squared or about
110 yards squared, the equivalent of about 2.5 acres). For weight, aside
from the familiar "kilogram" (about 2.2 pounds), the other measure recurring many times is the *quintal*, (the French plural is *quintaux*; and the
weight is the equivalent of 100 kilograms or 220 pounds). Metric and
American long tons are so close as not to require elaborate equivalents,
but when tons are mentioned in the text, they will always be metric. The
two most common (and synonymous) measures of volume in agriculture
are "sack" and "hectoliter," the equivalent of 100 liters or just under 3
bushels. A sack of wheat weighs or weighed (wheat is seldom sacked
these days) 80 kilograms more or less.

One final practical matter to put to rest concerns the nature of the
French administrative system largely inherited from the French Revolution. The lowest governmental entity is the *commune*, which I have
occasionally translated as "township" for want of a more precise equivalent. A *commune* may vary in acreage as in population: Paris is a *commune*, but so is Juzet d'Izaut, population 248. The *communes* that appear
in this book all happen to be villages with their surrounding countryside,
but the term's definition is political: it is the lowest unit of self-govern-
ment, electing about a dozen municipal councillors, who, in turn, select
the mayor from among themselves. In a fashion that Americans would
find very alien, a local mayor sees himself and is seen not only as the
political leader of his community, but also as the agent of the national
state in that community. Until recent laws that have decentralized the
system, he was clearly answerable to the high officials, prefect and sub-
prefects, who represented the state in, respectively, his *département*
(like Haute-Garonne, the size of three midwestern counties) and in its
subdivisions, the *arrondissements* or districts. The administrative link

between *commune* and *arrondissement* is the canton (normally a couple of dozen rural *communes*), the constituency electing a general councillor, who sits on the departmental general council, which until recently served as advisory body to the prefect, but which recent legislation has made into a full-fledged legislative body.

1

The Corncribs of Buzet

Two or three times the size of the area's average rural *commune*, Buzet-sur-Tarn is a sprawling township of 3,000 hectares (7,500 acres or about 12 square miles), located some 20 miles northeast of the city of Toulouse, just north of the *route nationale 88* that connects Toulouse to Albi.[1] In 1982, of a population of about 1,200, roughly one-third were retired people living in the village itself, one-third were commuters who, here and there, had built subdivision-style houses along rural roads and the last third were farm families, whose homesteads were scattered throughout the countryside. The main street of the village of Buzet is a departmental road that connects Saint Sulpice to the south to Villemur further north. Buzet village, to paraphrase the Michelin guides, is not worth a detour. With the shuttered look of so many villages of the French south, it is old without in any way being picturesque.

Yet Buzet does have its impressive landmarks that go unsung by the *Guide Michelin*. Visible from any of the *commune*'s backroads are historic monuments of a very special kind, the corncribs of Buzet. These are not your old-fashioned, wood-slatted, archetypal coffers of the American Great Depression era. Metal-framed and wire-meshed, Buzet's corncribs loom in the middle of some field, 20 feet tall and 300 to 400 feet long, looking like so many oversized, abandoned freight trains heaping full with hundreds of tons of yellow corn cobs—the land's bounty drying in the winter winds. This chapter is the story of how they got there.

We can take up the social and agricultural evolution of Buzet only from the eighteenth century, because only from that period on have such documents as tax rolls and land registers been preserved. By that time the village already had had five hundred stormy years of recorded chronicle behind it. It had served as a key fortress and administrative center of, initially, the counts of Toulouse, and, later, of the French kings, a fortress that kept changing hands throughout the Hundred Years' War of the fourteenth and fifteenth centuries, and the Wars of the Religion of the sixteenth.[2] Buzet seems to have been a genuine small town in the late Middle Ages but by the middle of the eighteenth century its population was only about 850, four-fifths of them peasants of one kind or another, most of whom probably lived in what was then still a walled village.[3] At least

that is where five-sixths of them could be found at the turn of the nineteenth century when the census asked for such information.[4]

Buzet in the eighteenth century was an agricultural community in
which rich outsiders occupied the top places of the socioeconomic hier-
archy. This clearly shows up in the distribution of landed property. According to the tax rolls of 1750, there were altogether seventeen properties above 20 hectares, (at a time when 12 hectares was considered the
upper limit for a manageable family farm).[5] Of the five top estates between 50 and 100 hectares, all but the smallest owned by a bourgeois,
were in noble hands. Of the twelve estates between 20 and 50 hectares,
three belonged to nobles, the rest to bourgeois. Altogether nine nobles,
mostly absentees (in the generation prior to the French Revolution, only
the Count de Clarac, Buzet's lord, was a very occasional resident), owned
25 percent of the total land; twenty-six bourgeois 35 percent of it. The
church, including individual clerics, had a 3 percent share. All of the
noble lands were let to tenant farmers on shares, as were most of the
bourgeois estates; church lands were normally leased for a fixed rent.
Only toward the very end of the eighteenth century did one or two landowners move toward managing their estates more directly by employing
families of farm servants known as *maîtres-valets*.[6]

Thirty-three percent of the land in Buzet was in the possession of two
hundred individual peasants (some of whom may have been members of
the same household) and an insignificant 1 percent belonged to seven
artisans or shopkeepers (of thirty-three households for that group). Buzet's peasants owned a larger share of the land than was the case in communities closer to Toulouse.[7] Evidently most of the estimated one hundred peasant households owned or rented some land, but only a small
minority had enough land to support a family. An informed guess would
place 10 percent of peasant owners in reasonably comfortable circumstances (eight peasants owned farms of 10 to 20 hectares and another two
just under 10); another 10 percent belonged to a "bare subsistence"
group (farms of 6 to 8 hectares); and a final 10 percent to a marginal group
that might make it in good years. Another 113 peasants owned an average
of 2 hectares each, about one-third of the vital minimum, and an additional sixty-four an average of three-quarters of a hectare.[8] There is no
way of knowing just how many of the local inhabitants were completely
landless, nor do we have any way of determining how many households
included several property owners. It is likely that at least three-quarters
of Buzet peasant households would not have had enough land of their
own to feed a family. Tenant farming must have provided openings for
thirty-five or forty families, the most fortunate of whom may have eaten
more regularly than proprietors of minifarms. This left another thirty-five

to forty households dependent on casual labor or on the rental of a field or two to supplement the harvest from their pathetic little plots.

If we are to make sense of what Buzet agriculture was up against, some understanding of the natural environment is essential. The climate of southwestern France has hot, dry summers, autumns that often turn rainy by November, a mild, wet winter with usually no more than a couple of weeks of frost, and uncertain weather in the early spring until summer effectively takes over in May. Though this is the typical pattern, the dry summer weather can readily turn to outright drought. Once every thirty or forty years, a winter may get so cold as to kill some of the grapevines and the mimosa trees, but there are few who have experienced this more than once or twice in their lifetime.

What does this mean for crops? The summer dry spell is always hard on perennial fodder crops like alfalfa, *luzerne*, clover, or grass and on late-maturing grains like corn. When the drought is prolonged, the effect is devastating. On the other hand, the intermittent rains during late fall and winter may interfere with the November corn harvest, with the plowing, harrowing, and sowing of winter wheat, indeed with the very survival of winter crops (young wheat survives two weeks' drowning, but may not survive four), and then again with preparing the soil for spring crops like corn. Until irrigation came along to change the rules, these had been facts of life for everyone working the land in Buzet.

The *commune* takes in four distinct subregions ordered by the course of the river Tarn and the layout of its valley. The river enters the township flowing almost straight west, but describes such a broad arc that it leaves Buzet almost on a northbound course. Some seven-eighths of the *commune* lies on the left bank, most of it quite level until you reach a low, knobby ridge roughly paralleling the river, but 2.5 to 4 kilometers distant from it. On the right bank there is only a narrow strip of level land never exceeding 500 meters, abruptly succeeded by an irregular line of bluffs, again vaguely parallel to the riverbed. These steep hills, rising to 300 or 400 feet, are indented by several narrow valleys.

The peculiarities of the soil are crucial. Sharp differences mark the two main agricultural plains within the township. The narrow strip of level land on the right bank is made up of a limestone clay soil that is regionally know as *terrefort*, the same "strong earth" that has made the nearby region of the Lauragais the traditional breadbasket of southern France. *Terrefort* is naturally fertile, but, above all, the clay stores rainwater, retaining it even during summer drought. This is a heavy soil that must always be plowed in the late summer or in the fall, even for crops that are not planted until spring. This has always been Buzet's best land, but there is not much of it.

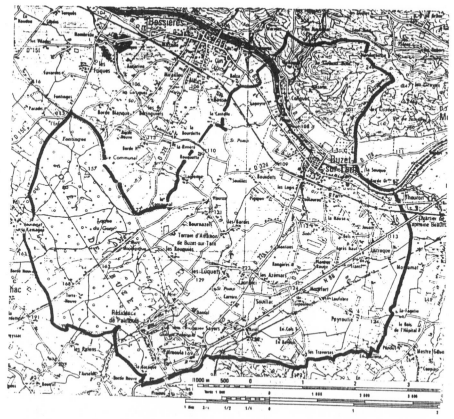

3. The township of Buzet

On the left bank of the river, the land is completely different. Around the village itself there is some rich alluvial soil, deposited by the flooding Tarn River or created by hundreds of years of putting dung and compost on kitchen gardens. Beyond the village perimeter, the greater part of the land is naturally poor soil known as *boulbène battante*, made up of finely grained clay fragments low on humus and nutrients, soil that the summer heat can turn to the consistency of adobe. Yet this type of soil remains soaked during the rainy season because under much of the land lies a layer of cement-like clay mixed with iron and manganese oxides, locally named *le grèpe*, that keeps rainwater trapped near the surface. Until very recently, a good winter rain could turn the plain of Buzet into a series of immense and long-standing puddles. For crops planted in the spring, *boulbène* must be plowed, harrowed, and sowed as soon as the March and April winds have sopped up any standing water and before the first heat dries it to unworkable adobe; for winter crops the soil must be prepared before the autumn rains turn the land to muck. At the same time, a sizeable portion of this left bank plain also consists of patches of gravel, a hectare here, a few dozen hectares there, several hundred all told. The gravel is even more bereft of topsoil than the *boulbène* and equally susceptible to drought, yet it also drains well. Poor for traditional cropping, worse for meadows, this gravel is naturally suited to vineyards.

Besides these two plains that have probably been farmed since time immemorial, the left bank is hedged by the large forest of Buzet implanted on soil that is hopeless for agriculture. On the right bank, the strip of fertile plain already described ends at the foot of steep, irregular bluffs, crowned by eroded hilltops that are today mostly scrubland and woods. Yet in contrast to the forest of Buzet on the other side of the valley, the hills on the right bank once served as Buzet's demographic safety valve. Many of the slopes are both sheltered and oriented toward the south or the southwest. The soil is sunny, fertile, and well drained. However, this is land so steep that it can only be worked by hand labor.

This unpromising agricultural setting accounts only in part for Buzet's evolution as a farming community. The fact is that, the intervening French Revolution notwithstanding, between the middle of the eighteenth and the middle of the nineteenth century, the peasants of Buzet increased in numbers, but lost out as property owners. Having slowly climbed to 1,000 inhabitants by the turn of the nineteenth century, in the next forty years Buzet reached a population figure of 1,320.[9] Where the peasants lost ground was in their share of the land. The largest domains became even larger, doubling in size and in the proportion of the *commune*'s land that they occupied. In 1750, estates above 50 hectares had taken up 25.3 percent of the *commune*'s total acreage; in 1838, 51 percent. Some of these gains came at the expense of smaller estates that had never been in peasant hands, yet peasant losses were also evident at every level: the acreage of substantial farms of 10 to 20 hectares, over half of which had been owned by peasants in 1750, declined by one-fifth; smaller farms of 5 to 10 hectares lost one percentage point; the total land in the possession of smallholders slid from 22.7 percent to 17.5 percent, even as their numbers exploded from 195 to 498, reflecting the pressures of a rising population.[10]

This dismal picture is darkened further by what we know of Buzet as an agricultural community during the 1830s. Much of Buzet's land was, as already noted, naturally infertile and hard to farm. Some 180 hectares, which coincides more or less with the soil surface that is gravel, was in grapevines, but their yield was only 10 hectoliters of wine per hectare, a small fraction of the average wine production of the 1980s. More generally, that same annual agricultural report of 1836, the only one to survive from the prerailway era, points to an agriculture that can only be described as medieval.[11] The ratio of seed to harvest for the two main crops, wheat and rye, were, respectively 1 to 5.5 and 1 to 4.4, proportions that could come right out of the thirteenth century. The net yield of a hectare of land planted to wheat averaged less than 550 kilograms.[12] Oats were even more catastrophic: the peasants of Buzet sowed one bushel of seed oats to gather 2.25 bushels at harvest time. Miserable as these yields are, they mislead by taking into account only land actually sown to grain. Figures for total acreage sown or in grass suggest that at any one time nearly half of the land was allowed to lie fallow. No summer crops of any sort were grown, save for 4 hectares of corn, presumably in the vicinity of the village where wells allowed watering. There was no clover or alfalfa or any other artificial meadows listed for the *commune*. Potatoes, later an important local food source, seem to have been unknown in the early nineteenth century, nor were enough green vegetables harvested to show up in the acreage statistics.

By any yardstick, Buzet was a desperately poor farming community that ate its home-grown rye and drank up two-thirds of its wine, while selling all of its wheat and one-third of its wine to outsiders. The total net value (that is, after deducting grain kept aside for next year's sowing) of the reported yearly cereal, wine, and meat production amounted to just under Fr 120,000. Even if all of the harvest had been evenly distributed among the peasant population, the per capita income would still have amounted to a measly Fr 150 per year. Since in real life much of the acreage was owned by absentee bourgeois and noble landowners who retained more than half of the harvest as their share, peasants' average per capita income must have been well below Fr 100 per year—at a time when the going *daily* wage of male farm hands in the nearby Lauragais was 75 centimes and a Parisian skilled worker could easily make Fr 75 a month.[13]

Yet Buzet's prospects turned out to be less predictably grim than these social and agricultural conditions portended. The evolution of Buzet's population provides some clues to unanticipated improvements. With the exception of market towns, most agricultural communities of the southwest reached a population peak some time in the middle of the nineteenth century, only to decline decade by decade thereafter. For example, nearby Loubens-Lauragais had halved its population between the 1850s and the outbreak of World War I. The demographic history of Buzet is sufficiently anomalous so as to hint at a genuinely divergent social development. In the case of Buzet, the population leveled off in the 1840s, remaining essentially stable over the next forty years. In 1881, it numbered 1,354. Then, over the next ten years, Buzet lost nearly a hundred people, dipping into the 1,100s before bouncing back to 1,255 on the eve of World War I.[14]

The reason why Buzet was able to retain most of its population down to 1914 had to do with the coming of the railroads, which opened up opportunities where none had existed before. In 1864, the line of the *Orléans et Midi* railway opened its station in Buzet; twenty years later, a second line, linking Montauban and Castres, also stopped at the village. Buzet had become a railroad junction.[15] The railroads became the only "outside" employer in the *commune*: by 1896 no less than fifteen families were directly dependent on railway employment.[16]

By 1885, when the schoolmaster of Buzet, like schoolmasters all over France, sat down to draw up a carefully structured report on "his" *commune* required of him by the minister of education, the railroads had revolutionized agriculture in Buzet.[17] Where in 1836 local farmers had planted 180 hectares in vineyards and drunk up two-thirds of the wine that they produced, by 1885 their vineyards took in 574 hectares, some of it on the gravel soil of the Tarn plain, some on sunny terraces carved out of the high bluffs on the river's right bank. Buzet's red wine was said to rate with Fronton and Villaudric among the best of the Haute-Garonne.[18] In contrast to 1836, Buzet now had seven times as much wine to sell and the ability to bring it to market conveniently and cheaply.

Even more novel was the move to take advantage of the sheltered, sunny little valleys and the southern exposure of the slopes of the right bank, where spring vegetables ripened two weeks earlier than anywhere else in the area. Anyone with a strong back and a plot of hillside land to turn over with spade or hoe could grow the green peas and string beans that for several decades made Buzet a byword for *primeurs*, that is, for early vegetables. Buzet came to have its own vegetable shippers and brokers, who, during the height of the season, saw to it that every day a freight car would be filled and expedited on its overnight trip to Paris. Some of the green peas of Buzet even reached London. In 1885 the writer of the official Buzet monograph estimated that the yearly vegetable crop alone was worth Fr 100,000—almost as much as the grand total of all of Buzet's products in 1836. The schoolmaster went on to explain the regrettable new trend toward small families as reflecting "the love of wealth developed by unrelenting labor." Elsewhere he off handedly explained that having school children held responsible for buying their own school supplies posed no local problems, because "there are no poor people in Buzet." In fifty years' time, a radical change for the better had taken place. The peasant community of the 1830s, seemingly crushed by nature, backwardness and social inequality, had been replaced by prosperous, market-oriented farmers.

Commercial wine and vegetable production were not the only achievements of the dynamic new Buzet. The new orientation toward specialized market crops relying on railway transport apparently also affected the traditional polyculture. Between 1836 and 1885 some of the land kept in fallow each year was put into artificial meadows, such as alfalfa, clover, and the like. The acreage planted to meadows expanded by a factor of eight at a time when there is no evidence that Buzet was increasing its livestock. Cows and oxen were still essentially considered work animals, rather than sources of milk and meat, though cows did have the good sense to drop a yearly calf. The presence of a local "fodder broker" on the census list of 1896 suggests that hay was now being grown for sale and shipment in the same way as wine and green vegetables. Wheat had completely displaced rye, the poor man's bread, which had practically disappeared. There is also some evidence that agricultural technology was beginning to change: the two *entrepreneurs* (contractors) listed in the same census were almost certainly harvesting contractors running the threshing machines that first appeared in the French southwest in the last quarter of the nineteenth century.

Yet the abrupt decline of Buzet's population by over 10 percent in the fifteen years between 1881 and 1896 was a symptom that something had gone awry with the village's agricultural prosperity. In all likelihood, the emigrants of the 1880s and 1890s were vintners or their hired help ruined by phylloxera, the grapevine infestation that had reached Europe from North America and was rapidly wiping out France's (and Europe's) vineyards in the last three decades of the nineteenth century. Phylloxera had first reached Buzet's vines in 1881.[19] Though the deadly aphids spread slowly at first, the outcome—the destruction of the existing vineyards— was, as elsewhere, only a matter of time. Not until the late 1890s, when the technique of grafting French vines on American-French hybrid rootstocks that resisted aphids yet suited French soils triumphed over competing palliatives, was there any chance of recovery from the disaster.[20] Replanting, however, was a very costly business and after 1900, low wine prices, due partly to Algerian competition, made it prohibitive in many marginal wine areas like Buzet. By 1913, long after phylloxera had been overcome, Buzet was left with 170 hectares in vineyards, less than one-third of its acreage of 1885.[21] The local vineyards never recovered; over the next forty years, the amount of land in grapes hovered at about the 1913 level.[22] Most farmers continued to have a hectare or two of grapes and to make and sell their wine, but few of them produced enough wine to rely on it alone for their living.

Even so, Buzet's prewar agriculture seems to have largely succeeded in circumventing the effects of the vineyards' decline. Despite continued low yields—only about 25 percent more wheat per hectare was produced in 1913 (a mediocre year, to be sure) than in 1836—wheat remained king.

No agricultural revolution was to be detected there. Yet Buzet farmers did spectacularly better with barley as well as with oats, traditionally a crop for poor soils, though the official 1913 figures look almost too good to be true. Potatoes and Jerusalem artichokes had also become significant crops. Some 400 hectares, over one-fourth of all cropland, were now in artificial meadows producing fodder and perhaps commercial seed. If the statistics can be trusted, the age-old practice of leaving the land lie fallow as part of the normal crop rotation—that is, leaving each field plowed, but unsown and unplanted, one year out of two or three—was rapidly falling into disuse.[23] One indicator of relative prosperity was the high ratio of oxen, the draft animal of choice, to cows, the draft animal of necessity.[24] Yet above all, what seems to have kept Buzet going despite the plight of the vineyards was the continued high production of peas and beans, two-thirds of them green, one-third dried. Almost 100 hectares were planted to these two crops, a considerable acreage for what really amounted to a form of intensive truck gardening.

Buzet's evolution was more typical in the unprecedented impact that World War I had on its population. From 1911 to 1921, population dropped from 1,255 to 912, a decrease of over one quarter. The downward slide continued more gently between the wars, reaching an all-time low of 784 in 1946. To account for this sharp decline, the ubiquitous World War I village war memorial tells only part of the story. For every name on the marble plaque, there was at least one other veteran who came back maimed and unable to farm. Some of these disabled veterans were likely to move to cities like Toulouse, where they could hope to become state-licensed tobacconists or be offered a job in the post office. As one older Buzet farmer told me, only three or four of those drafted in 1916 (the year his own father's turn came) ever went back to farming after the war had ended. In fact, the continued population slide between the two wars was steeper than what the official statistics indicate. From the mid-1920s on, rural demographic statistics no longer measure the same population as previously: the indigenous population declined more sharply than the census figures indicate, for successive waves of outsiders filled some of the vacant places. During the 1920s and 1930s, North Italian sharecroppers passed through Buzet, a few of them ultimately settling permanently, followed by a wavelet of Spanish refugee farmhands, none of whom stayed for good, but both groups bolstered census figures.

The shock of World War I and the decline in Buzet's population to which it led seems to have affected all but the vineyards, already reduced by phylloxera earlier. By 1921, Buzet's specialty crop, peas and beans, green and dried, was down to 55 hectares, a drop of 45 percent since 1913. It had gone down another few hectares by 1932.[25] In 1913 grain crops had occupied 785 hectares, in 1921 370 hectares, in 1932 just over 420. In 1913 fodder crops took up 400 hectares, in 1932 just over 250. In fact, of 1,570 hectares of arable land listed, the crop statistics of 1932 account for less than 750 hectares. In 1913 they had accounted for about 1,400. This was an indication that either fallowing had staged a comeback or, more likely, that a considerable acreage had gone out of cultivation altogether. This is suggested by other figures as well: between 1913 and 1932, the number of work-oxen went down from 260 to 72, the number of work-cows from 330 to 314 and this at a time when local tractors could be counted on two fingers. Between the wars, agriculture was in obvious decline in Buzet.

This cannot be described as a case of the young leaving Buzet because some technological revolution had displaced them. Between the two wars, the gardeners high up on the bluffs, sowing their beans on Saint Joseph's Day (19 March), were all men getting on in years. The younger generation, survivors of the war, would have none of this back-breaking labor that had sustained Buzet for two generations. As the old gardeners died off, no one took their place and their carefully tended plots reverted to what they had once been—thickets. Besides, even earlier produce, grown under easier conditions, was now reaching Toulouse and Paris from further south, mostly from the Perpignan area in French Catalonia. Produce dealers stopped coming to the thinning Buzet weekly market, although two local shippers hung on throughout the 1930s. In 1940, the line of demarcation drawn by the Nazis, cutting off occupied from unoccupied France, sundered what was left of Buzet's famous green peas from their Parisian market.

Although truck gardening was a long time in dying, the basic physiognomy of traditional peasant farming remained unaltered. During the interwar period, a good part of Buzet's acreage was still owned and controlled by absentees who let their land in family-sized farms of 10 to 20 hectares to sharecroppers, *maîtres-valets*, and, on rare occasions, to tenant farmers paying a fixed rent in money. Almost one-third of the land was still in estates of over 35 hectares. Between 1914 and 1938, not much change had occurred in the number of farms between 5 and 20 hectares, nor in the land that they held. Where there was considerable change was at the very bottom of the social scale: the number of proprietors of less than 5 hectares was halved, from 685 to 365, but their claim to Buzet's farmland went from 23.3 percent to 29.2 percent of the total acreage.[26] In 1936, when the census happened to be exceptionally specific, forty-one farmers were described as owner-operators, including a half dozen miscellaneous gardeners, nursery owners, and horticulturists; forty-one as sharecroppers; and four as fixed-lease tenant farmers. In addition, forty-five individuals were included in the catch-all category of *cultiva-*

teur, which seems to have included married peasant sons living with their parents, individuals who combined ownership with sharecropping, owners of plots of less than 1 hectare (bean and pea growers?), and possibly some *maîtres-valets*, the group intermediate between sharecroppers and farm servants.[27]

Family polyculture was once again the rule, specialization the exception. In 1929, only one single farm of 4 hectares claimed to be entirely devoted to grape-growing.[28] An unaided family with two male adults—and by this time there were only eleven hired hands and five agricultural day laborers left within the *commune*—could manage only 15 to 20 hectares at the most (38 to 50 acres). Such a farm produced wheat for bread and cash, fodder, barley and oats for the work animals (in contrast to the prewar period, now more likely to be cows than oxen), occasionally some clover or alfalfa seed, a calf or two a year, a bit of corn for the few dozen fowl and the yearly pig, and two hectares' worth of wine, most of it for sale. The slow pace of plowing with cattle and the limited time in which that task had to be completed made larger family farms impractical.

The basic pattern of peasant farming did not change during the interwar era, but some of the peripheral techniques did. Though mechanization did not cause the flight from the countryside, the reverse is true: between the two wars the shortage of harvest workers certainly encouraged mechanization. Although this was a quieter, cheaper, less traumatic process than the breakneck modernization that engulfed the rural south beginning in the 1950s, this earlier change was nonetheless both significant and widespread. The agricultural inquiry of 1929, for example, found fifty double Brabant plows in Buzet. The double Brabant is an excellent wheeled plow, perfected in the mid–nineteenth century, that can be set to plow to any depth (provided the traction is powerful enough) without human pressure having to be applied to keep the plowshare cutting the furrow evenly. Of course, these figures turned upside down also meant that at the onset of the second quarter of the twentieth century, more than half the farmers of Buzet may still have been using the traditional Roman *araire*, which lacking a true mouldboard, literally scratched the surface, unable to make a furrow deeper than five inches, although there were also intermediate plows around that were less primitive. Sickle and scythe for harvesting grain had been pretty well displaced: there were seventy simple reapers in the *commune* and twenty of the more sophisticated reaper-binders that tied the wheat or barley stalks into sheaves. The hay harvest no longer required repeated hand raking either: there were seventy so-called horse-rakes and ten more elaborate hay-tossing machines. The single threshing machine listed undoubtedly belonged to the harvesting contractor, as must have one or two hay and straw presses (yet there were six listed for Buzet), a couple of tractors, and the plows with multiple shares to match. Most likely, Buzet's one and only mechanical seed-drill, its single disk-harrow, its sole fertilizer spreader, its solitary mobile urea dispenser—equipment that by local standards was thirty years ahead of its time—belonged to some improving landlord, perhaps to the owner of the community's largest estate, the Chateau of Conques.

The positive significance of this incomplete, yet widely diffused, mechanization of the interwar years is clear. Most of these technological changes were geared to saving on harvest labor, labor that in the nineteenth century had been recruited from among the numerous *brassiers*, the farm laborers whose arms were for hire. Significantly, this was a social class, now relabeled day workers (*journaliers*) or farmworkers (*ouvriers agricoles*) that by 1936 had shrunk to a mere five families.[29] In that sense, the mechanization of harvesting (as the earlier mechanization of threshing) permitted peasants on family farms to break a labor bottleneck that might have prevented them from carrying on as in the past. This was therefore a conservative modernization allowing the small family farm to survive without its traditional complement of casual labor at the peak season of the year.

The limitations of these changes are equally clear. They did not speed up plowing, the major constraint that limited the size of the family farm. The double Brabant was undoubtedly more efficient and less fatiguing than the almost prehistoric *araire*, but it was still being pulled by the same pair of oxen or team of work-cows at exactly the same slow pace. Mechanization was intended to eliminate hired harvest hands; it was not intended to raise the labor productivity of farm operators in their other tasks. Nor does there seem to have been any urgent concern with agricultural productivity as such. Old farmers reminiscing about the introduction of the Brabant plow still remain more impressed with its ability to guide itself, than with the agronomic effects of deeper plowing. Artificial fertilizer was known and used, mostly in the form of "superphosphates," but in minute quantities broadcast by hand like seed and then only on crops like alfalfa, clover, and corn. No one aped the unidentified owner of the 1929 mechanical seed-drill and fertilizer spreader, both of them tools that were to revolutionize grain yields starting in the 1960s. Moreover, the machinery most peasant farmers acquired at this time was relatively simple and quite cheap. The purchase of a horse-drawn hay-rake did not call for private soul searching or a fundamental reevaluation of the entire farm operation. It was the kind of equipment that most peasants could afford without changing their lives.

In the long run, World War II galvanized France into an effort at transformation and modernization that was bound to alter Buzet's evolution as

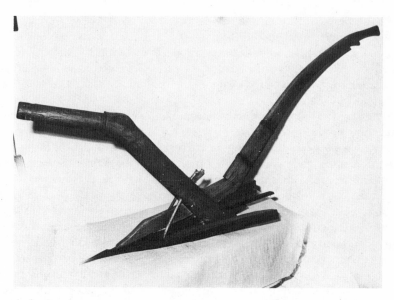

4. *Araire*

5. Double Brabant plow

6. Plowing with an *araire* (one team of oxen)

7. Plowing with a double Brabant (three teams of oxen)

an agricultural community. Yet this did not become evident until the 1950s. By contrast, the immediate effects of the war were obvious. A notable minority of peasants—the national figure is 13 percent of adult male farmers, better than one out of eight—spent the war years as German captives.[30] They are thought to have returned in 1945 more open to new agricultural ideas after spending the war working on German farms technically more advanced than those that they had left in the French southwest. Some of the smallest farmers left in Buzet were given the choice of being deported to work in German war industries or to take a job in the coal mines of Carmaux. As already noted, the German occupation severed Buzet's peas and beans from its Parisian markets, a disruption from which local truck gardening never really recovered.

On a day-by-day basis, compared to most other villages in the southwest, Buzet had a hard time under the German occupation. In 1944 the area was occupied by a Waffen S.S. division, with a garrison in Buzet to guard the railroad station. Trigger-happy troops enforced a 9:00 P.M. curfew, occasionally leaving victims by the side of the road and, incidentally, forcing farmers to curtail their harvest workdays. If we can believe reminiscences forty years after the events, Buzet lived out the war in an atmosphere of pervasive fear and, as it turned out, with good reason: in the summer of 1944, the mayor and a number of local young men were rounded up by Nazi raiding parties in the dead of night and summarily executed as members of the resistance.

Yet when the war was over and the prisoners of war returned home to Buzet, nothing much seemed to have changed. On the right bank of the Tarn, the noble owners of the Chateau of Conques continued to farm their domain with *maîtres-valets*, the traditional system of hiring servants with their families. More land in the *commune* continued to be sharecropped than was farmed by peasant owners. Around the village proper and up on the slopes of the right bank, the aging survivors of Buzet's truck farming went back to hoeing their peas and beans. For the rest of the agricultural community, the traditional pattern of local polyculture had never been interrupted: a little bit of this and that to feed the family, some fodder and oats to take care of the work animals, some *luzerne* grown for seed, a calf or two a year to sell on the market at Bessières or Saint Sulpice, and a couple of hectares of grapes with wine as the chief cash crop. Nor had much changed in the way of farm equipment since the agricultural inquiry of 1929, except that the harvester-binder had by now become generally accepted among the larger farms. The state-supported farmers' bank, the Crédit Agricole, still played a minimal role. The only recourse to a loan that was considered legitimate was on the occasion of taking over the parental farm, a solemn moment that required paying off one's siblings, who were entitled to their share of the inheritance or to its equivalent in the form of annual payments, the so-called *soulte*.

The prevailing field system suited this predominantly subsistence agriculture: small family farms, whether owned by the operators or share-cropped, usually had a large number of scattered strips, many of them of a size that could be plowed with cows or oxen in the course of a morning or a day. The fact that they were scattered may have been a random result of the ebb and flow of family inheritances, yet such dispersion also provided the only practical insurance against hail, a natural mishap that was often highly localized, destroying crops in one field, while leaving untouched fields 100 yards away. Moreover, scattered fields allowed those peasants prosperous enough to own a shotgun the pleasure of stalking hares all over the *commune* without leaving their own land.

This seeming immobility of the agricultural community disguised major changes in the making. The garden culture around the village and on the hillsides that had sustained the descendants of the day laborers for three generations finally flickered out as the last truck gardeners died off. The reputation of Buzet spring peas died with them and by the early 1950s their land, which had once commanded premium prices, had become tracts of prickly wilderness. It marked the end of the era during which Buzet's farmers had successfully capitalized on railway transport.

The rural exodus that had drained Buzet of one-third of its population since World War I also assumed new dimensions. Toulouse had traditionally served as a regional magnet for rural immigrants, but after World War II the pace of immigration accelerated, as the city entered a phase of extraordinary dynamism and expansion. For the first time, even if they had no special skills, the young in Buzet could look for work in Toulouse and be practically sure of finding it. By the 1950s it had become obvious that becoming a farmer or a farmer's wife was no longer a matter of fate, but of choice. And at a time when continuing to live in a peasant household in Buzet still meant putting up with drawing water from a well for the woman and walking fifteen or twenty miles a day behind an ox-drawn plow for the man, city living seemed clearly a step up, *une promotion sociale*, to employ the untranslatable French jargon. Nor was there any pervasive regional mystique about the supposed superiority of rural life to counteract the attraction of the city. The older generation was at best ambivalent: they needed their offspring to carry on, yet they themselves had no illusions about the harshness and penury of traditional peasant life in Buzet. In many cases, parents, particularly mothers, seem to have encouraged their own sons and daughters to seek an easier life in the city.

Yet this rural mass exodus, which all local people mention as a fact of life, cannot be documented from the official population statistics of the post–World War II era. On the contrary, between 1946 and 1954 the number of inhabitants crept from 784 to 818. Thereafter the population appears to have risen steadily to 877 in 1962, to 1,060 in 1975, and to nearly 1,200 by the end of 1982. The reason for this anomaly is that the outflow of discontented farmers, farmers' sons, and farmers' daughters was more than offset by an inflow of new people into the *commune*. A minority of the newcomers became part of Buzet's farming community: between the end of World War II and the 1960s. Buzet experienced the influx of half a dozen Aveyron dairymen (all of whom bought property and remained), of a number of equally stable farm families of North Italian origin (who had reached France in the 1920s or 1930s) who began their French careers as hired hands or sharecroppers, and French expatriates from North Africa, some of whom proved more volatile. The net effect of this migration was that by the 1980s, among farmers less than sixty years old, indigenous Buzetois had become a minority. Yet a *majority* of the newcomers had nothing whatever to do with agriculture. A large number were natives who had left for the city, returning to their village, *au pays*, as the French say, only for their retirement. A second large group consisted of commuters employed in Toulouse, but sometimes with family links to Buzet, who, unable to afford inflated Toulouse housing prices, put up their homes wherever they could manage to buy a building lot. In sum, the links between population and agriculture became increasingly tenuous.

The dynamic expansion of nearby Toulouse with its seemingly unlimited opportunities for unskilled or semiskilled labor in the building trades was but one of the factors that helped to undermine sharecropping, the traditional southwestern system of tenant farming, which, as we have seen, accounted for more than half of the farms of any size in Buzet. If there was little glory in inheriting a parental farm that was family property, the prospects of taking over as a sharecropper were even bleaker. Peasant proprietors had to contend with harsh working conditions, recalcitrant soil and weather, and unpredictable markets. Sharecroppers not only faced the same problems, but everything they produced was divided fifty-fifty with an owner who was also unmistakably *le patron*, the boss. Sharecropping was not only a lean living, but it was demeaning, and, since leases ran only from one year to the next, humiliatingly insecure. Unlike peasant proprietors who had to overcome their traditional attachment to a family holding before making a fresh start in the city, sharecroppers, not to mention their children, had no such sentimental ties to discard.

If urban full employment undercut sharecropping, it was also challenged by restrictive legislation and sapped by economic changes that made traditional estates hiring sharecroppers or *maîtres-valets* unprofitable. Since the disintegration of the traditional estate system based on sharecropping played a far more central role in Loubens than it did in Buzet, we shall put off its discussion until we can deal with it more fully in the context of Loubens-Lauragais. Suffice it to say that in Buzet, the Chateau of Conques, the last great aristocratic agricultural domain, began selling its lands, the best in the *commune*, by the late 1940s. By the beginning of the 1960s all of them were gone, mostly transformed into American-style orchards owned and managed by repatriated Frenchmen from North Africa. Other traditional landowners followed suit. A good many farms that had long been sharecropped were put on the market by their disgruntled owners. Between 1950 and the late 1960s property changed hands on a scale probably never seen before. Into the late 1950s it remained a buyer's market, until the massive land purchases by North African repatriates pushed prices up, persuading many remaining absentee owners that land still had potential as an investment. Tenancy also shifted from sharecropping to long-term fixed leases, a process that took about twenty-five years to work itself out in Buzet. By the 1970s sharecroppers had become an anomaly; by 1982 they were extinct.

At the same time, a combination of factors—the juridical procedure by which the range of rents for farm leases was set, the spectacular increases in land prices, and the no less spectacular revolution in crop yields from 1960 on—turned tenant farming topsy-turvy. The departmental tribunal of tenants and landlords routinely fixed yearly rents in terms of 400 to 600 kilograms of wheat, a grain whose yearly price increase failed to keep pace with the general price index.[31] In real terms, land leases became cheaper year by year. By 1982, in Buzet lease payments averaged Fr 600 per year per hectare on land where irrigation could produce Fr 12,000 worth of corn. The tenants' rent amounted to about 6 percent of the gross product and 9 to 12 percent of the net. As by then land prices had risen to Fr 25,000 and Fr 30,000 per hectare, from the landowner's perspective a net of Fr 560 (the proprietor, not the tenant, was saddled with the land tax) amounted to a return of a mere 2 percent on his landed capital. From the vantage point of the farmer-operator, this made leasing land a far more economical method of enlarging one's holding than acquiring that land by purchase. For progressive young farmers the ancient mystique of land ownership had become irrelevant. "We are here to make money from farming," one of the second-generation Italian dairy farmers explained deftly. "We are not here to spend our money so we can end up farming" (*On est plutôt là pour faire de l'argent avec l'agriculture que de faire de l'agriculture avec l'argent*). Finally, some proprietors of land became increasingly reluctant to lease it out officially, not so much because of low returns, but because long-term formal leases prevented them from cashing in on the rising market in building lots. Consequently, a shadowy amount of Buzet land was rented out under the counter, so to speak, to

tenant farmers willing to work it on a year-by-year basis without any written lease.

If there ever was such a thing as a symbolic turning point, a peasant's
first purchase of a tractor ought to qualify as such. Aside from a few pioneers and a handful of stragglers, farmers in Buzet motorized in the
course of a dozen years between 1950 and 1962. The crest of the wave
was between 1952 and 1956, when most of those buying acquired hand-
me-downs from northern France, tiny tractors of 15 or 20 horsepower,
often equipped with gasoline engines, tractors that proved to be so uneconomical to run that they had to be traded in within a short time. Since
most of these starter tractors were inexpensive to begin with and second
hand to boot, they could be purchased by selling a pair of oxen or some
work cows. For the second tractor, likely to have a little more horsepower and a diesel engine, many farmers took out the first equipment
loan at the Crédit Agricole—a radical departure both for the peasants
concerned and for the authorized farmers' bank. At the time of the agricultural census of 1954, about one out of four of the serious farms in
Buzet owned a tractor.[32] By the mid-1960s even marginal farms of 7 hectares were under overwhelming pressure to motorize: by then trained
work cows—and most farmers were not used to training them themselves—were impossible to find. By the 1970s tourists driving through
Buzet stopped traffic gawking at the last old farmer who, to the day he
died, continued plowing with cows. By that time another census had
counted ninety-five tractors for the *commune*'s sixty-three farms: draft
animals had become *du folklore*.[33]

Although the timing of motorization in the Buzet farming community
can be determined with reasonable accuracy, it is much harder to find
out why local farmers made this momentous move when they did. Buzet
farmers merely explain that this was the time when everybody was buying a tractor, an explanation that is suspiciously circular. No doubt, even
in a rural society that is often characterized as intensely individualistic,
imitation plays a much larger role than is usually assumed. ("Farming is
easy," one Buzetois assured me. "All you have to do is watch what all
your neighbors are doing and go out and do it yourself.") Once tractors
had become common, those not yet motorized were indeed likely to follow. Yet tractors were not some odd novelty first seen in 1950: harvesting
contractors in Buzet had used them since the 1920s without inciting imitation.

The real answer probably cannot be found at the local level. The shock
of military catastrophe and occupation had instilled a will to national renewal, which in the postwar years was most often translated into an un-
critical faith in technological progress. That was a national phenomenon

in which the peasants of the French southwest shared. They read the same *Depêche de Toulouse* and listened to the same national broadcasts in the postwar years as anyone else. The pervasive neopositivist message was widely internalized. "You have to keep up with progress" (*il faut suivre l'évolution*) became a veritable litany among French farmers of the southwest, a litany I heard repeated dozens of times.

As far as French farming was concerned, the new ethos was translated into both plans and policy. The first national economic plan, formulated in 1946, called for agricultural motorization, though the particular implementation it envisaged—a sort of French version of the Soviet tractor stations—was misguided and soon abandoned as impractical. From that point on, the government encouraged the mass production and mass import of low-horsepower, inexpensive tractors suited to small-scale polyculture, tractors the like of which had never been widely available in the country.[34] Furthermore, within a few years, with guidance from on high, the semipublic Crédit Agricole reversed its traditionally restrictive farm lending policy in order to encourage the acquisition of the new equipment. By the early 1950s the more progressive northern farmers were ready to trade in their first postwar tractors for heavier equipment, creating a national secondhand market in small tractors in the process and a massive geographical shift of farm equipment from north to south. In at least that sense Buzet's peasants bought tractors when they did "because they were there."

Students of traditional peasant cultures have long known that such peasant societies have their own economic rationale, but that they possess only limited notions of economic rationality as market economists define it.[35] Certainly when the opportunity arose, Buzet peasants had not been adverse to making money by selling their wine and their early vegetables, but their primary concern was to feed their families. Near self-sufficiency in homegrown food, supplemented by some cash crops, was the ideal. The accumulation of capital posed no problems for an established farm, equipment being simple, cheap, and durable; draft animals were self-renewing; trained oxen were bought young, fattened, and sold at a profit a few years later; work-cows netted a yearly dividend in the form of a calf before they were ultimately sold to the butcher. Like the peasant household itself, work-animals lived off the land, which provided the pasture, fodder, and oats that they needed.

How then did the purchase of a tractor disrupt this equilibrium? First of all, it forced its owner to look at his farm with new eyes, as a minicapitalist farmer rather than as a peasant. The tractor was not just another useful tool as the reaper had been. This struck home most forcibly when a peasant had bought his tractor by borrowing money and knew that for five years on such and such a date he would have so much principal and interest to repay. Yet that was only part of the story. Even if he had been able to pay cash for his first tractor, a tractor cost so much a year in fuel and repairs, and, even more important, had to be amortized in a way that cows and oxen had not. "Tractors don't have young," is the way one retired Buzet farmer put it wryly. In short, tractors required farms that were sufficiently profitable—in cash and not simply in family food resources—to run, maintain, and replace tractors, and ever bigger and costlier tractors at that. Tractors exuded the notion of profitability (*rent-abilité*) as unavoidably as they exuded exhaust fumes. And once this alien notion took hold, the question was less whether a 10 or 15 hectare farm could afford a tractor, but whether you could justify a 10 or 15 hectare farm once you had a tractor. The tractor became the Trojan horse introducing a novel economic rationality that was to disrupt the traditional framework of the family farm.

In the intermediate run, as an expensive capital asset to be amortized, the tractor called for a major enlargement of that family farm. As a technical innovation, the tractor also made such an enlargement practical. The time required for plowing had always limited the maximum practical size of a farm and tractors turned out to be incredible time-savers. Given the fact that much of Buzet's soil was unworkable throughout the winter because it turned to bog and to be equally unworkable as soon as the first heat wave baked it, plowing during the brief interval when reasonable soil conditions prevailed was a natural bottleneck. Tractors broke that bottleneck or at least widened the neck of the bottle. Compared to an oxen- or cow-drawn plow, even a 20 horsepower tractor dragging a single-share plow took only one-third as long to work the same field and that was only a preview of things to come.

This was the point where three distinct trends happened to converge: the technical ability to save time in the plowing; the need to utilize the new tractors as fully as possible in order to spread the new overhead costs as widely as practical; land becoming available, both because old peasant proprietors retired or died without direct succession or without heirs willing to come back to farming from the city, and because formerly sharecropped farms were made available for rental, lease, or sale. The outcome was a gradual, yet in the end radical, increase in the size of those farms having either an assured succession or else young people already at the helm, at the expense of farm operations having neither. Farms smaller than 10 hectares were usually doomed; some 15-hectare farms enlarged, others were absorbed. The determining variable was a son's or son's-in-law inclination or disinclination to stay on the farm.

This process appears less nebulous if we turn to a specific example by tracing the expansion over time of one farm operation that local people would hold up as both well run and "normal." The Cadoulenc family came to buy a farm in Buzet in 1947, after their own dairy farm was flooded by the construction of a dam in their native Aveyron, a hilly *département* a hundred some miles to the northeast. If there is anything atypical in their story, it would be the fact that theirs became the first and, for a number of years, the only, dairy farm in Buzet and that the 25 hectares purchased were acreage enough to justify a small tractor without financial strain. They had bought what had once been one of the outfarms of the Château of Conques on the right bank of the Tarn, which gave them the double advantage of excellent *terrefort* soil and a single, consolidated block of land convenient for pasturing cattle. Cadoulenc Senior bought their first 15 horsepower tractor in 1952 and a more powerful one of 35 horsepower with a two-share plow in 1966, which allowed the son, while still working with his father, to rent a 12 hectare farm on the left bank of the river. The father officially retired in 1969, though for the next sixteen years he remained in charge of the gradually diminishing herd of milk cows. The son began by taking out a preferential start-up loan for young farmers in order to buy more powerful equipment. He then leased still another farm of 12 hectares to help amortize that equipment, bringing his total holdings to nearly 50 hectares. In the 1970s Cadoulenc picked up three more small farms on the left bank, whose owners retired or died. Two of these were leased officially, but one continued to be rented by verbal agreement only. Cadoulenc, all by himself now, worked 76 hectares—an acreage 25 percent larger than the traditional American quarter section of 160 acres—yet he *owned* no more land than his father did in 1947. By his own account and by that of his neighbors, 76 hectares, almost half of it irrigated, were about as much as one man could handle. What made it possible at all was that the soil on the two opposite sides of the Tarn was plowed at different seasons of the year and that Cadoulenc had three working tractors at his disposal, the newest and most powerful of which could cut four furrows at a time.

Not only did the tractor provide both the incentive and the possibility for farm expansion, but almost immediately it led to a drive to make the farm, such as it was, more profitable by intensifying its operation. The first response was to turn to dairy farming; a second, somewhat more delayed, reaction was to attempt increasing crop yields by using modern equipment and employing modern methods to which local people had previously paid little or no attention. Though other considerations also fostered change, the tractor seems to have been its catalyst.

In the course of the 1950s what had been an exotic import from Aveyron—dairying—became the norm among practically all Buzet farms above 10 hectares and on many that were smaller. Prior to the advent of the tractor, every farm of substance had had its work animals, usually a pair of oxen and a pair of cows, sometimes only four work-cows, rarely a total of as many as six adult animals. The oxen were usually sold as soon as a tractor was bought, leaving what cows remained good for nothing except calving and for providing manure.

Here again was a situation where many things came together. The cows remaining may have been underemployed, but their dung was essential to the agricultural cycle. The tractor saved the farmer's time, thereby making time for a new activity like milking. The tractor also called for higher revenues from the farm operations and milk was a major boost over merely producing calves. Furthermore, on inferior Buzet soils, grass and fodder crops did much better than winter crops like wheat and infinitely better than summer crops like corn. In some cases, the expansion of a farm triggered by the tractor also made up for the dispersion of the fields originally attached to that farm: it was difficult to have even a modest herd of cattle unless there was a solid block of pasture land adjoining the cow barn. In some cases renting land from retiring neighbors made such pastures possible for the first time.

The net result of all this was that milking herds appeared, sometimes simply by farmers keeping old work-stock for a new purpose and raising heifers they previously would have sold. Occasionally some of the more prosperous and enterprising went out and bought themselves a herd of as many as ten dairy cows. By 1960 practically everybody in Buzet was milking cows, though they rarely kept more than one cow for every 2 hectares of land. Dairying did not replace polyculture: it merely added yet another facet to an agriculture marked by diversity.

For most farmers there seems to have been a considerable delay between the time when they acquired their first tractor (followed soon afterwards by plows and harrows that were carried on the tractor, rather than having to be dragged behind) and serious attempts to raise the productivity of the land by the use of more effective methods and the equipment for which these called. By that time, most farmers were on their second generation of tractors: the gasoline-powered toys had all been replaced by more powerful machines, often with 35 horsepower diesel engines. There were no longer any oxen to trade in for that diesel tractor, so most farmers were in hock to the Crédit Agricole, a situation that would have increased their receptivity to innovations that promised to increase their cash income. By this time, not only had the Crédit Agricole become extraordinarily responsive to underwriting productive loans for farmers, but experience and inflation were eroding the age-old inhibition against borrowing. As in the case of acquiring a first tractor, there must have been some spillover from national trends: during the early 1960s France was experiencing booming industrial growth. Dynamism and progress were in, caution and tradition were out.

What made these attitudes relevant was the availability of information and advice. By the end of the 1950s a network of agricultural extension agents, one per canton, had been put into place in Haute-Garonne. New regulations required young farmers to obtain a minimum of professional schooling, which most of that generation obtained by attending night or weekend courses taught by extension agents, if they wanted to benefit from preferential equipment loans when they took over the family farm. In at least one instance, a retired farmer maintained that his son, now considered among the most progressive younger farmers in Buzet, was first converted to a scientific, high-yield agriculture after reluctantly attending an adult agricultural education program.

Finally, the very fact that in Buzet there was a substantial nucleus of farmers' sons who chose farming as a career at a time when the rural exodus had become a flood also affected the pace of change, though probably not its direction. The older generation was willing to bend over backward to keep its sons interested in taking over the family farm and the young took full advantage of their enhanced bargaining power. This induced fathers to go along with the methods pushed by their sons. Farmers who had no succession had no real incentive to switch to promising, but untried and expensive agricultural techniques. Even today when modern agriculture has proved itself in Buzet, farmers over fifty years of age whose succession is not assured consistently refuse to invest or even to renew depreciated equipment.

What made the Buzet situation different from other agricultural communities in the *département* was the degree of cohesion that these up-and-coming farmers achieved. Even though most of them had ended their formal education with at most the *certificat d'études*, the terminal school certificate that the best school leavers are awarded at the age of about thirteen, they knew at least the direction that they wanted to take and, from the early 1960s on, seem to have been aware that their cooperation enhanced chances of reaching their goals. Technically untrained, at the beginning they had a clearer notion of their aims than of the means by which they could be achieved. Their underlying commitment was to remain as farmers, but they also looked forward to running progressive farms (*des exploitations dynamiques*) large enough to employ modern equipment and profitable enough to provide a middle-class standard of living—and this at a time when the average southwestern farmer made less than a *smigard*, the minimum wage earner. In the late 1960s they were joined by outsiders who acquired farms in Buzet, a few more Aveyronnais, some second-generation Italians, an odd Breton. A few years older than the Buzet young bloods, the newcomers latched on to their group, not only because they shared its aspirations, but because they saw it as an oasis in what they deplored as Buzet's apathy and backwardness. Nationally, this heady brew of agrarianism, productivism, and the mystique of entrepreneurship—the young farmer as *chef d'entre-prise*, as CEO—was identified with the most dynamic of France's post-war agricultural organizations, the *Cercle des Jeunes Agriculteurs*.[36] Not surprisingly, its local chapter provided an ideal organizational framework for Buzet's cohort of young farmers.

In retrospect, there is wide agreement that effective leadership had a great deal to do with the agricultural upswing of Buzet. Enter Mallorca, a young *rapatrié* from Algeria, whose father and cousin at the beginning of the 1960s bought the major part of the estate of the old Chateau of Conques on the right bank of the Tarn, gradually transforming it into apple orchards. Mallorca had attended and graduated as an agronomic engineer from the Ecole de Purpan, the only university-level agricultural school in the Toulouse area. He was no meteor, he explained to me, hav-ing come to Buzet to stay, yet he also refused to spend his life in a dying village. A man of great energy and evident competence, Mallorca won over his Buzet contemporaries by force of character, rather than by charm. With the devastating assurance that even minor French elite schools—Purpan, a Catholic, rather than a state institution is at best a marginal *grande école*—instill in their students and alumni, Mallorca proceeded to change the milieu by taking charge of the Cercle des Jeunes Agriculteurs, of which he became president.

Though he may have seen himself as a potter at work, Mallorca was as successful as he was because he was handed unusually malleable clay. The members of Cercle were themselves committed young farmers, able and determined to master their trade. Mallorca's ideal, farmers who took charge of their own lives, individually and collectively, neatly dovetailed with his Young Farmers' own views of farming as a chosen career rather than as a hereditary disability. His training had given him some perspec-tive on what his followers needed to learn for themselves and middle-class status had taught him the administrative ropes. Yet though he some-times undoubtedly rubbed people the wrong way—modesty was not his forte—he was genuinely concerned that his members should learn both to cooperate and to think for themselves. It was the members of the Cer-cle des Jeunes Agriculteurs who decided what they ought to take up and in what order. Over the years they became the prize pupils of the agri-cultural extension service, arranging for workshops and training sessions on everything from irrigation to cost accounting. In the 1970s a nucleus of them even managed to find financing for a study trip to the Province of Québec from which they returned with the satisfaction of knowing that the Québecois had little to teach them. Along the way, the new genera-tion of Buzet farmers gradually achieved an extraordinary degree of com-petence in every practical, technical, and managerial aspect of farming.

As should be expected, the agricultural technicians responded enthusiastically to people who welcomed what they had to give.

Where Mallorca's personal conviction seems to have had most influence on his followers was in his championship of what the French call *agriculture de groupe*, feebly translated as collective, but not collectivist, agriculture. He believed in the free association of small nuclei of farmers in order to overcome the disabilities that have traditionally burdened farmers' lives: the inability to take vacations; the risk of illness disrupting an ongoing farm operation. French farm legislation of the 1960s had encouraged such ventures by providing a legal framework (with significant tax and credit advantages to match), the GAEC (Groupement agricole d'économie en commun, which may be rendered as "joint agricultural venture").[37] Mallorca himself, with his father and cousin, had formed the first such GAEC in Buzet and one of the first in the Haute-Garonne. The formula has usually, though not necessarily, linked associates who are related by blood or marriage, a GAEC having the advantage of spelling out responsibilities, rights, and benefits for all concerned in ways that traditional family arrangements do not. Psychologically, it was associated with a rational, up-to-date approach to farming itself and to the division of labor and profits among the partners in a collective agricultural venture. In most *communes* of Haute-Garonne finding even one GAEC is uncommon; Buzet is unique in having more than half a dozen.

The Cercle des Jeunes Agriculteurs may have had only an indirect effect in promoting the initial technical changes that during the 1960s began creating high-investment, high-yield agriculture in Buzet. In any case, not every innovation fitted that mold: the first, and, after the tractor itself, the most impressive technical novelty was the triumph, between 1955 and 1963, of the harvester-combine, which cut the grain, threshed it, bagged it (a practice later abandoned when bulk transport and storage proved less costly), and, on some models, baled the straw as well. On the American prairie the machine had been in universal use for two or three generations, yet it reached the French southwest only in the early 1950s. As late as 1956, a local harvest contractor had still bought a new and improved model of the old-style threshing machine in the mistaken belief that the combine was not really here to stay. By 1963, he bowed to the inevitable after barely depreciating his thresher suddenly gone obsolete. This is one case where a machine was not directly related to the push for greater intensification, but was more akin to the old reapers, reaper-binders, and mechanical hayrakes—harvest labor-saving devices all. And until well into the 1970s it was beyond the individual means of even the most prosperous Buzet farmers, much as threshing machines had also been owned by harvest contractors and never by ordinary peasants.

That the new combines saved an immense amount of very hard work cannot be questioned. The old system of harvesting grain was by means of a reaper-binder usually pulled by oxen, sometimes by two yoke of oxen, because the machine required considerable effort. The machine cut the grain, tied the stalks (in more advanced models automatically) into sheaves, and dropped these sheaves onto the ground as it proceeded around the field. The sheaves had to be piled on carts, wagons, or (after the end of World War II) farm trailers to be brought into the farmyard, where the family, usually helped by neighbors, systematically stacked these sheaves into huge pointed strawricks, looking like so many gigantic, upright straw cannon shells 20 or 25 feet tall. The grain-bearing kernels of the sheaves all pointed toward the interior of the rick, for it was absolutely crucial that the wheat, oats, or barley kernels remain dry for the days or weeks that it took before the threshing machine put in its appearance. As soon as the machine did reach the neighborhood, the same group of a couple of dozen neighbors, known regionally as *la colle*, followed it and its hired crew from farm to farm for two or three weeks until all the local threshing was completed and a *colle* from an adjoining section of the township took over. At each farm the women not only participated in the work, but prepared the gargantuan meal or meals according to which the host family was going to be judged. The work itself was horrendously hot, dusty, itchy, dirty, noisy, and smelly, but this mutual aid did provide a sense of community that is still fondly remembered by those getting on in years. Yet before our eyes turn moist, we should recall the plight of our Buzet harvesting contractor: when farmers had an alternative before them back in the late 1950s they massively chose the antisocial combine over the jolly old threshing machine. In so doing, they not only chose ease and progress—*il faut suivre l'évolution*—but equity as well: small farmers had long complained that the system of the *colle* was unjust. Even though they had much less to thresh than their more prosperous neighbors, they put in just as much time as the big farmers.

In Buzet the new techniques geared to a new productivist agriculture appeared a little later than the combine—between 1960 and 1965—and they appeared as a cluster. By this time the great cooperatives, many of them going back to the interwar period, had established their predominance over private grain and seed dealers. In the case of Buzet, it was L'Occitane to whom farmers sold their wheat, barley, and corn and from whom they bought their supplies, seed included.

As already noted, since the 1920s or even earlier small quantities of artifical fertilizers had been applied to fodder crops and, less consistently, to corn. Nowadays the newly developed and more productive varieties of seed grain pushed by the coops require heavy doses of such fertilizers, though at this point these were supplemented by the increased dung that dairy farming had made available. One farmer specified that his consumption of artificial fertilizer rose from 1 ton to 11 tons between the late 1950s and 1982, although his increased acreage may have accounted for some of this higher consumption. Fertilizers in large quantities called for fertilizer spreaders, first of the type dispersing solid pellets in the manner of salt trucks servicing snow-covered highways, later traded in for pressure sprayers designed to cover a swath as wide as 25 feet with liquid chemicals. These latter machines could also double for spraying herbicides and fungicides, products that came into massive use during the 1970s.

As part of the same productivist revolution, during the early 1960s the seed-drill universally replaced the age-old method of broadcasting seed by hand. Again it is striking that we are dealing with a technology developed in the eighteenth century and perfected in the first half of the nineteenth. The 1929 agricultural census had counted a single seed-drill in Buzet. The seed-drill's victory of the 1960s reflected a change of outlook triggered by the tractor—or else the tractor itself was already a symptom as much as a cause. In any event, the advantages of the seed-drill over broadcasting seed were decisive: only one-third to one-half as much seed was required by the new method, while the even rows of grain produced by the seed-drill grew better than wheat sown at random.

The result of this new agriculture was a rapid increase in grain production per hectare that more than repaid the new investments. It did not transform poor grain land—a description that fits most of Buzet's land on the left bank of the Tarn—into excellent land. Only the wheat and barley that survived excessive winter rains benefited, as did the corn that managed to live through summer drought without being drowned at harvest time. The new agriculture was likely to make the new breed of mechanized, progressive young farmers more, rather than less, resentful at seeing their disciplined efforts come up against limits that previously had simply been accepted as fate.

The young farmers of the Cercle des Jeunes Agriculteurs led by Mallorca provided the major impetus for what proved to be a decisive new stage in the process of agricultural modernization. It was they who agitated for land consolidation, *remembrement* in French, a term that suggests a body whose members are being joined together again. *Remembrement* was the government-sponsored process of redrawing field boundaries within a community in such a way as to assemble agricultural land around each home farm in the form of large fields suitable for modern farm equipment. Since the end of World War II, French law had encouraged this procedure as part of the general national effort of making French agriculture more modern and more productive.[38]

This encouragement took the form of the state financing the entire process of surveying and reallocating farm land and heavily subsidizing (with the *département* of Haute-Garonne chipping in additionally) related public works, what in the jargon of the civil engineers are called *les travaux connexes*, that are just as important as redrawing the fields as such. These related public works involve digging drainage ditches, bulldozing superfluous hedges and boundary ridges, and improving both field access roads and rural communal roads as well. Above all, the *remembrement* of a township was an extraordinarily complicated multilateral exchange of land designed to create more conveniently laid-out farms. Official land consolidation was initiated only upon the demand of the mayor and municipal council of a *commune*, who, the departmental officials of the Ministry of Agriculture assumed, spoke for the wishes of the local farmers.

The fact was that Buzet *was* severely dismembered. As has been noted earlier, fields were scattered all over and very small in size. As we saw, this dispersion had served as a sort of informal hail insurance, yet by the 1960s formal hail insurance had become virtually universal, as it was required by the Crédit Agricole before any loan was approved. The young farmers who were pushing for land consolidation had three things in mind. First, they wanted larger fields that took modern equipment into consideration. A one-acre field (about Buzet's average) was uneconomical to work with a powerful tractor and a nightmare for a large harvester-combine, particularly if the field was in the form of a long, narrow strip. Even if it could be managed, there was a tremendous waste of time involved: it was infinitely faster plowing a single fifteen-acre field than fifteen scattered one-acre strips. Second, irrigation required cohesive properties, because having three or four neighbors' fields scrambled among one's own made laying out a rational irrigation network virtually impossible. Third, early individual efforts at irrigation had shown that in solving the problem of summer drought, extended watering aggravated the problem of winter drowning. Land consolidation promised government-subsidized and thereby inexpensive drainage ditches to relieve the problem of excess water.

Getting the ball rolling turned out to be quite easy. Mallorca and his Jeunes Agriculteurs convinced the mayor, a young man who was not himself a farmer, but who happened to be the nephew by marriage of the president of the regional Chambre d'Agriculture, the semipublic bureaucracy that administered the myriad of services that farmers have come to take for granted. As a local politician he was certainly eager to support a project advertised as essential to keeping the local young on the farm. The municipal council, comprising members of both the younger and the older generation, proved equally sympathetic. Public meetings, attended by as many as one hundred landowners, also revealed majority support for initiating land consolidation. The fact that Haute-Garonne was way

behind many other *départements* in the number of requests for land consolidation made the regional officials of the Ministry of Agriculture also
highly receptive. On 25 April 1966, the municipal council of Buzet-sur-
Tarn officially requested the prefect of Haute-Garonne to set the process
of land consolidation in motion. The *remembrement* of Buzet was
launched, not to be completed until December 1970, more than four and
a half years later.[39]

The details of the ponderous procedure by which *remembrement* normally proceeds need not detain us. Like any government-sponsored operation, land consolidation has its regular phases in what may be compared to a bureaucratic minuet. Successively, *remembrement* deals with
the specific acreage to be included in the perimeter of land to be consolidated; the classification, that is, the qualitative assessment, of existing
fields; the preliminary and then the definitive proposals for the new field
system; the carrying out of the related public works projects previously
decided upon; and, finally, the assumption by each landowner of the land
assigned or reassigned to him or her. Each of the early stages is followed
by an elaborate inquiry at which local landowners are free to voice their
objections and to try to redress their grievances. If they fail to obtain
satisfaction informally, they may appeal, first to a communal, then to a
departmental commission. In extreme cases, a plaintiff may, even without hiring a lawyer, carry his case to an administrative judge and ultimately even up to the Council of State in Paris, the highest administrative tribunal in the land.

One of the problems that made land consolidation unusually difficult
at Buzet was the incredibly large number of landowners, no less than 476
of them at the time that the process began. Some of the larger absentee
owners had sold out to working farmers by the beginning of the 1960s,
when land prices shot up two- to fourfold, responding to the sudden demand for farm land by tens of thousands of displaced French farmers
from North Africa. By contrast, owners of 1 or 2 hectares—most of them
ex-farmers or their heirs who had gone to Toulouse—generally held on
to their real estate in the hope that by and by it could be used or sold as
building lots, a hope that the growing number of suburbanites in Buzet
encouraged.

Both the sheer number of proprietors and their nonagricultural aspirations exacerbated the built-in dilemmas of French legislation on land
consolidation. Such legislation had a fundamental inherent contradiction:
its goal was to improve the structure of French farms, many of them increasingly based on leased rather than on operator-owned land, yet the
actual process by which land was consolidated was on the basis of *ownership*, rather than of *land utilized*.

In retrospect everyone agrees that one major aspect of this problem could have been readily solved by setting aside a building reserve near the village, a reserve in which any landowner who so desired could have been assigned building lots. Apparently Mallorca, a member of the communal land consolidation commission, proposed such a scheme, but the proposal was dismissed on the grounds that it complicated further a process already extraordinarily complex.

The solution that was favored was shortsighted: to reassign plots of land belonging to smallholders anywhere along existing rural roads, both in order to unclutter the area most suitable to farming and to provide potential building lots for these small owners. The results were predictable: on one hand, uncontrolled and scattered residential building that was bound to strain the municipal budget for public services; on the other hand, conflict between resident farmers and resident commuters, who had not bargained for noisy tractors and smelly spray fertilizers in what they saw as their own backyard. When ten years later, local farmers very belatedly insisted on zoning as a way of keeping outsiders away from farming areas, the small landowners, to whom land consolidation had assigned plots now suddenly barred for housing construction, felt understandably cheated. In this sense, *remembrement* left an uncomfortable legacy of divisive political conflict.

In Buzet, aside from this problem of land speculation, the large number of proprietors did set limits to the degree of land consolidation that was practical. Prior to 1970, sixty-five active farmers had worked 1,460 hectares divided into 1,114 blocks of land (the number of distinct fields recognized in the official land register being about twice that number). Land consolidation reduced these separate chunks of farm land to 365, no mean feat, yet this was a far cry from *remembrement*'s ideal of granting to each farmer one cohesive block of land around his residence.[40]

As was the case with most land consolidations, not everything was sweetness and light. Older farmers who did not have sons to push them deeply resented having their lives disrupted. They were likely to sign petitions indicating their extreme chagrin. The aggressive and articulate, moreover, fared better than the timid and the reticent when it came to the assessing, classifying, and swapping of fields. It was also true that dairy farmers who lost established meadowland in this redrawing of the local map or farmers who had sulphur and copper-soaked former vineyard land foisted on them were handicapped for several years afterwards. Nonetheless, with few exceptions, most Buzet farmers still active twelve years later agreed that only land consolidation had made further agricultural progress possible and that, without it, many of the local young farmers would be working in the city today. As a knowledgeable and otherwise very matter-of-fact harvesting contractor put it, "*remembrement* has been this area's good fortune."

This process was much less successful in remedying the problem of waterlogged winter fields. In this case, the *Direction départementale de l'Agriculture* (DDA) and its chosen surveyor proposed public works, but it was the local commission of landowners created for the occasion that decided how much was going to be spent. Buzet farmers opted for doing things on the cheap: they were willing to spend Fr 10 per hectare per year for twenty years and not one centime more. It should have caused no surprise that the drainage ditches dug turned out to be totally inadequate: ten years later they had to be deepened at four times the original cost.

Land consolidation did have one major unanticipated side effect: it accelerated the decline of the Buzet vineyards that had begun with the phylloxera infestation of the 1880s and 1890s. From 160 hectares in 1954, grapes went to 85 by 1970–1971 after land consolidation had been completed.[41] One of the early decisions of the communal commission in charge of *remembrement* was to allow any owner of a vineyard or an orchard to retain it if he or she so chose. This occasionally led to such quirks as the case of an old peasant who, in the midst of somebody else's field, managed to cling to his four rows of asparagus by having them declared an orchard. What was less expected, this liberal approach decided most farmers whose vineyard was distant from their farm residence to trade it for land closer at hand, where they had the option to replant grapes. Yet many chose not to exercise this option, but to choose the alternative of selling their right to plant grapevines, planting rights for which there is a strange, yet publicly recognized, nationwide market. The result was that Buzet's vineyards shrank markedly in the course of the four and half years it took for *remembrement* to be completed.

Yet land consolidation was only one significant factor in hastening the end of the winegrowing era in Buzet's agricultural history. Aside from the shock of the phylloxera epidemic that cut its vineyards to a third of the area that they had occupied previously, a further decline began to be felt shortly after the end of World War II. It is true that at the time there still survived swarms of peasants with tiny holdings, who managed to hang on because they worked a couple of hectares of vines. These were precisely the model unpromising small farms whose heirs were most likely to flee to the city, leaving aging parents to tend, prune, spray, and harvest. Then in 1956 southwestern France was hit by the kind of big freeze that saw temperatures plummeting to −20° Fahrenheit, a disaster which seems to occur once every generation. A substantial portion of the Buzet vineyards froze and died. Replanting involved not only a heavy investment in time and money, but it required a wait of at least five years before the vines would once more become productive. Many marginal farmers, unable to afford the investment or the waiting, either quit, or, if they had sufficient land, switched to other crops. Unfortunately for the future of winegrowing in Buzet, those who did replant tended to follow official advice and put in hybrid plants that produced generous quantities of rot-gut wine. Just when these vines began to bear by the early 1960s the market for *gros rouge*, the ordinary red that had sustained generations of French vintners, began to decline. What also played against Buzet wine was the mechanized bent of the younger generation. Many of the surviving vineyards had so little space between rows that even the narrow, specialized vintner's tractors could not pass. It was bad enough that cutting back the vines during the winter months was strictly hand labor, but the inability to use *any* mechanical equipment made vineyards an even less attractive proposition for the young. By the time official land consolidation got started, all of these factors—the 1956 freeze, the marginality of most wine-oriented farms and their lack of succession, the inability to mechanize grape growing, and the relative stagnation of wine prices—had already begun to change the picture.

The termination of land consolidation also marked a changing of the guard among Buzet's farmers. The generation that had gone along with *remembrement* in response to their sons' insistence now turned over their farms to them. By the early 1970s retirement at age sixty-five had as yet not become second nature among farmers, the basic social security legislation for farmers dating only from 1955. Over and above a minimum pension paid to all farmers and to their spouses upon reaching the age of retirement, the amount of the old-age pension varied with the number of years social security payments had been made. Collecting old-age pension payments did not require turning over one's farm to a younger farmer, but ever since 1962, a substantial supplementary pension for life, the so-called IVD (*Indemnité viagère de départ*—roughly, lifetime compensation for quitting) was available if one did.[42] Though they turned over responsibility for the family farm to their son, most parents continued to lend a hand.

Of the two key problems facing most of Buzet's farmers—excessively wet soil throughout the winter and lack of adequate rainfall during the summer—the problem of summer drought began to be addressed as early as 1963. Around the village there had always been wells by means of which people watered their gardens, but the pioneer in installing a modern irrigation system was young Cadoulenc on the right bank of the Tarn River. There was a certain irony to this, since at this time all of the Cadoulenc land consisted of *terrefort*, a soil that resisted drought better than any other and that had therefore less to gain from irrigation. When the son persuaded his father to invest in irrigation, having been unable at this time to find any additional land to lease, he saw it as an alternative to expanding the acreage farmed. This itself shows the close links between intensifying agriculture and expanding acreage. Since the Cadoulenc property fronted on the Tarn River, water was no problem and by the early 1960s the first effective sprinkler systems were becoming commercially available in the southwest. In contrast to most farms in Buzet prior to *remembrement*, the Cadoulenc farm consisted of a single block of land, which facilitated setting up an irrigation system.

Most other young farmers could not begin irrigating until after the completion of *remembrement* and until they were in charge of the family farm and able to obtain the favorable credit terms granted to qualified young farmers about to launch on their own. This was the case of the Grasouillet farm, where the son began irrigating after taking over from his father in 1970 and leasing additional land. Unlike other farmers in his neighborhood, who despite considerable effort and expense failed to find adequate aquifers, Grasouillet found water on his land at 5 meters, digging a number of wells. Here again, irrigation was too expensive an operation to undertake casually: Grasouillet had long wanted to sell his father's dairy herd, because "he and cows just never got along." Specializing in irrigated corn would allow him to make up for the income lost from milk. By the early 1970s a minority of farmers were irrigating some of their land to grow corn, a crop that had been notoriously risky in Buzet. The solution was not wholly satisfactory even for those lucky enough to strike water on their property: the wells were generally inadequate as to water output and subject to drying up when they were most needed— during unusually parched summers.

The conviction that irrigation was indeed a solution to a major part of their problems, but that individual irrigation systems were out of the question for some of them and unreliable for others, induced the Young Turks of Buzet, still under Mallorca's leadership, to seek a collective approach: a collective, but not a political approach, in that the Jeunes Agriculteurs did not want to be tied to some official, municipally controlled arrangement. Typically in Haute-Garonne irrigation tends to be highly politicized in that the formal initiative usually comes from the mayor of a village, who will then negotiate with the requisite hydraulic service of the DDA in Toulouse or, in the case of communities in the northern quadrant of the *département*, deal with the Compagnie des Côteaux de Gascogne, a regional nonprofit corporation specializing in irrigation projects. In such cases, farmers can sign up or not as they see fit, but the irrigation association remains a municipal enterprise largely administered from above and backed by the municipality's credit.

At Buzet it was not the municipality dispensing the manna, but the farmers organizing to obtain irrigation on their own initiative that constituted such a departure from the regional norm. The juridical instrument that Mallorca and his adherents selected was based on nineteenth-century laws dating back to 1865 and 1884, authorizing public recognition for an organization of landowners planning to undertake major collective improvements. Normally this involved a common agreement, drawn up free of charge, before the mayor, who then certified that each signer of the instrument was indeed the proprietor of specified parcels of land within the township. After publication of an official announcement in a local newspaper, what amounted to the papers of incorporation were then officially filed with the prefect, an act that conveyed semiofficial standing to the association.

The Buzet group, rejecting this routine approach, decided on a much more unusual procedure to symbolize the intensity of their commitment to their joint venture. They chose to file before a *notaire*—that unique Roman-law equivalent of a British solicitor, an American registrar of deeds, and a French land transfer tax collector all rolled into one. By doing so, they solemnly and jointly committed all of their real property as collateral against any future outlays for irrigation in Buzet.[43] Local people were in awe that, in order to make a symbolic point, these young farmers had been willing to pay the hefty notarial fee for something that they could have done for free. Unfortunately such formalities have no resonance for those of us whom the French insist on calling *Anglo-Saxons*, but in Buzet they impressed everyone, the participants included, with the fact that the new irrigation association meant business.

They received a lot of help with their planning. The section chief of the agricultural extension agents, a specialist in regional irrigation problems who had worked extensively with the Cercle des Jeunes Agriculteurs, undertook a feasibility study on behalf of the Chambre d'Agriculture. His conclusions were favorable. The irrigation service of the DDA provided free advice. Interns from the Agricultural School of Purpan, Mallorca's alma mater, volunteered, attracted by Buzet's growing reputation as a hive of go-getters. The regional agricultural management center, also sponsored by the Chambre d'Agriculture, examined Buzet's problems in order to determine what crops would make the most of irrigated land. As though it were short of help, the Buzet Irrigation Association also retained a firm of consulting civil engineers who had made a regional name for themselves by planning and supervising a municipal irrigation scheme at nearby Villemur-sur-Tarn.

In the interim much practical preparation had to be completed. Every farmer in Buzet was personally approached about joining the association, not once, but several times, and given to understand that once the irrigation network was in place, there would be no opportunity for joining later. The drought of 1976, which revealed the shortcomings of individual irrigation, helped reinforce their arguments. Before the books were closed on new members, twenty-five farmers with a combined total acreage of 900 hectares (roughly two-thirds of Buzet's arable land) had joined the irrigation association. Most members intended to water about half of their land at any one time.

There were a variety of reasons why some farmers turned down collective irrigation. Theoretically, the cost was incumbent on the owners rather than on the tenants of landed property. In real life, if land was under proper agricultural lease, proprietors merely gave their consent, but it was the tenants who assumed the expense, something they were generally willing to do provided they enjoyed security of tenure. Tenants renting land (illegally) by oral agreement on a year-to-year basis therefore could not afford to take such a chance. There also were still occasional farmers who simply hated to borrow, particularly if they were making only a marginal living and lacked confidence in their ability to repay a long-term loan. In some cases, people were deterred by happenstance: one of the local harvesting contractors doubling as farmer turned down irrigation because technically his recently inherited farm was *en indivis*, that is, still jointly owned with his siblings whom he knew to be unwilling to invest in improvements. Cadoulenc, for his part, saw no reason for changing an irrigation system that was already fully amortized, cheap to operate, and which had worked adequately in the driest of summers. Yet, for the most part, the abstainers were small farmers over fifty years of age, who, having no successor, felt that they could not justify heavy, long-term indebtedness.

Mallorca was given to lapidary formulas about the need for modern farmers to stand on their own two feet, instead of scrounging for subsidies and ending up as professional bounty hunters, as *chasseurs de primes*. Ironically, its rhetoric notwithstanding, subsidies were what the Buzet Irrigation Association was all about. From its inception in 1973, the new irrigation association made major efforts to obtain official funding that would allow it to proceed with constructing a pumping station on the Tarn and laying their irrigation pipes. Encouraged by a former classmate who headed the irrigation section of the DDA in Toulouse, Mallorca proposed that the Buzet Irrigation Association apply to the European Economic Community (EEC) for support. By the EEC's criteria, the French southwest qualified as an agriculturally underdeveloped area. There was no reason, so the argument ran, why Italian farmers should be the only ones to garner support payments. If no French farmer had ever received such funding, it was because they had never bothered presenting a cogent case.

The campaign ran into trouble from the start, as it ran counter to strong national prejudices. President De Gaulle had emphasized tough bargaining over farm prices within the EEC, but had discouraged other sorts of EEC influence over French agriculture as a breach of national sovereignty. This tradition of holding aloof had persisted even after De Gaulle's departure, impeding the Buzet farmers' approach to the EEC. It is true that Mallorca, who had political ambitions, had recently become chairman of a national association of GAEC's, which gave him some visibility and influence. He was, however, clearly a man of the political Right, which may have been a plus in dealing with the Ministry of Agriculture in Paris, but his unequivocal political stance made him *persona non grata* with the Socialist office-holders dominating the Chambre d'Agriculture of Haute-Garonne. It so happened that, as official representative of the organized profession, support from the regional Chambre d'Agriculture carried decisive weight and these officials were not going to bestir themselves for one of their fiercest and most vocal critics.

It was probably in part because the members of the Buzet Irrigation Association found themselves in this limbo of red tape, that a core of five of them, still headed by Mallorca, embarked on an experiment in group agriculture by founding a GAEC, that is, a cooperative enterprise, of their own to raise young steer. Some of its members had already begun irrigating individually. All of them were eager to find an outlet for their own corn at prices higher than those they received from the coop. By transforming corn into beef, they reasoned, they should be able to sell their corn to their own GAEC at near retail prices and make an additional profit selling fattened bullocks. By creating a cooperative partnership in the form of a GAEC, their commitment would be limited in scope and not directly involve their own family farms. Above all, they had no intention of recreating what they saw as the slavery of dairy farming: partnership was a way of making sure that no one would be excessively tied down. At the same time, the new GAEC justified considerable investment in heavy equipment, such as a harvester-combine owned jointly that they could never have afforded individually.

With help from the Crédit Agricole and a regional cooperative that promised to buy their bullocks ready for market, the five launched their venture. They rented land, bought and put up sheds, purchased the requisite farm equipment. A little over one hundred male calves made up the stock. The first year they did brilliantly, buying cheap and selling dear. Unfortunately for them, a slump in beef prices intervened and, unlike most cattle raisers, they kept elaborate analytical accounts that left them with no illusions: while they were not going broke, if one took labor costs into account (and they did keep track of their own hours as part of the bookkeeping) as well as return on capital invested, they turned out to be no better off than when they had sold their corn directly to the coop. Mallorca was the first to withdraw. By 1979 the cattle-raising

GAEC disbanded altogether, its combine transferred to an equipment coop, a CUMA (*Coopérative d'utilisation de matériel agricole* or coop for the use of farm equipment), created for the occasion. By that time, in any case, the Buzet Irrigation Association once more held center stage.

As there have always been avenues for bypassing the entrenched French bureaucracy, so there should have been ways of getting around EEC officialdom in Brussels. And apparently there were. The key to any wire pulling was the young mayor of Buzet, whose uncle-by-marriage had meanwhile shown the acumen of getting himself appointed vice-president to one of the EEC's agricultural commissions. Unfortunately for the business of the Irrigation Association, however, by the 1970s Mallorca and the ex-mayor (who had just completed his term) were on a political collision course that had nothing to do with the agricultural future of Buzet. Both were ambitious; both were competing for the same elective office; neither was willing to compromise. Personal political rivalry ended up short-circuiting the Buzet Irrigation Association's quest for EEC funding. Not long afterwards, Mallorca withdrew from the Irrigation Association's chairmanship and from his role as champion of Buzet's modernization. Moving up to loftier circles, he was elected president of several elite agricultural organizations before becoming president of the French apple-growers' association, as well as founder and director of its ultramodern, regional cooperative packing house. It was an ironic tribute to the quality of his leadership that his withdrawal from local affairs made no perceptible difference in Buzet. His ambition had been to train young farmers who would be able to cope and cope they did. For the time being, they gave up on the EEC, turning to more familiar and, as it turned out, more responsive agencies for support. By 1979, the Buzet Irrigation Association was given the green light by the Ministry of Agriculture, which agreed to pay 60 percent of the cost of the permanent irrigation installation, with the *département* of Haute-Garonne forking over 20 percent of the remainder.[44]

In fact, by the time collective irrigation finally received its go-ahead, local farmers had already garnered considerable experience with individual irrigation. In this respect, Cadoulenc's was wholly favorable, but he did have unusual advantages not shared by others; the *terrefort* soil on his home farm required only limited watering and the land he rented on the left bank just happened to consist of well-drained gravel. In these conditions, without encountering any special problems, he regularly produced 9,000 kilograms (dry-weight) of corn per hectare on land that had previously yielded 5,000 kilograms in a good year and 2,500 kilograms in a year of drought.

The experience of another early irrigator, Grasouillet, was more typical. With the wells that he had dug, he was able to irrigate 25 hectares, using a system known as "total cover," which is total only if each and every one of the sprinklers is moved by hand every eight hours. In 1971, when he began, this was the only type of low-volume sprinkler system available. For four years, Grasouillet grew crop after crop of irrigated corn on the same land, pleased with his excellent yields and delighted to be able to amortize his equipment so quickly. In retrospect, he realized that some of his excellent results should have been credited to the fact that his land was practically virgin as far as intensive agriculture was concerned. Nineteen seventy-four turned out to be an exceedingly wet year, to the point where the late autumn fields turned into quagmires. Only combines equipped with caterpillar treads could harvest what turned out to be yet another bumper corn crop. The heavy equipment on the soggy soil ruined the fields, leaving deep ruts and compacted earth.

The following year, 1975, was even more humid than the year preceding. It was becoming obvious that the combination of irrigated single cropping, heavy harvesting equipment, and unusually rainy weather had altered the structure of the soil. The earth gave off a foul stink when it was being plowed, as though it were rotting or mildewed. In fact, the soil had become waterlogged, asphyxiated to the point where roots had trouble growing, a problem with which even major increases in chemical fertilizers could not cope. This condition showed up in patches of spoiled soil where nothing whatever would now grow. It showed up more widely in the increasing difficulty of preparing the land for crops. Fields that had been plowed by a 60 horsepower tractor mounting triple plowshares now required tractors with twice that horsepower. Even fields that a generation earlier had been plowed with wooden *araires* would now only yield to heavy equipment. And heavier equipment evidently tended to compact the soil still further, thus perpetuating the vicious cycle. And, of course, the inherent economic logic of the tractor continued to prevail: the heavier and more expensive the equipment, the more acreage was needed to amortize and justify it. In Grasouillet's case, he justified a new four-wheel-drive tractor by leasing another 15 hectares.

By the end of the 1970s, at the very moment when collective irrigation had finally been funded and was ready to start up, the risks of high-intensity monoculture had become obvious. Grasouillet's experience and that of others was also showing that it was much easier to damage the structure of the soil than to restore it. Soil could deteriorate dramatically within two or three years, but might take a decade or more of unrelenting effort to regenerate. Dairy farmers growing irrigated corn for silage were not as affected: their corn was harvested much earlier, before the fall rains had added to the water problem caused by summer irrigation. The fact that their corn was immediately followed by a winter cover of Italian ryegrass that fixed nitrogen in the soil also helped. Grain farmers, with varying degrees of reluctance, had come around again to some form of regular crop rotation to avoid long-term disaster. The alternation of soybeans (irrigated, but much less intensively than corn) and corn have become the recent standard. Some farmers have come to consider variants of the traditional three-crop rotation with nonirrigated wheat or barley to relieve the stress of excessive watering.

The more acute Buzet farmers recognized early on that returning to some form of crop rotation was essential, but not sufficient. Work methods also had to change, particularly at a time when heavier four-wheel-drive tractors could get through under any soil and weather conditions and the new rotative harrows could break up clods no matter what. Despite the time pressure imposed by increasing acreage, farmers came to make an effort to avoid plowing wet fields, even though their increasingly powerful machines could have passed even then. They doubled tractor tires in order to spread the weight. They deliberately reduced the number of times a tractor would pass back and forth over a field by increasing the effective width of their spray.

At the same time, agricultural extension agents were trying to convince Buzet farmers that something more fundamental was needed if they wanted to maintain the initial magnificent yields that irrigation had permitted. Only drainage, the technicians argued, could rid their land of the excess water the irrigation sprinklers unloosed all summer upon a soil that already had a natural tendency to waterlogging in the winter. Before irrigation, dry summers had helped the land to recuperate from the wet winters. Originally their suggestion had raised serious questions as to whether the soil predominating in the *commune* of Buzet, *boulbène battante*, could be improved by drainage at all. Test-tube waving scientists of some repute had insisted that it could not, until, in fine disregard of their findings, farmers in various parts of France tried draining *boulbène* soils with splendid results. Maquille, the extension agent who had worked most closely with the Buzet team on a host of problems, including his specialty, irrigation, took his charges on a demonstration trip to the adjoining *département* of Gers to persuade them that drainage could do wonders for waterlogged *boulbène* soils. The soil of the village in Gers and that of Buzet turned out to be demonstrably similar, but the additional expense that drainage involved seemed prohibitive at a time when the members of the Buzet Irrigation Association were just about to start paying for their collective watering network.

There was nothing newfangled about drainage, since the ancient Romans had already drained some of their farm land and drainage had been widely used in northern France down to World War I. For reasons that remain unclear, not much draining took place between the two wars, but the practice revived in the 1960s with modernized techniques. The principle was elementary, however complicated the actual execution of a large drainage project. It involved burying networks of pipes, with rows of holes in their top half, at a depth safe from plowshares and, like underground gutters, laid at a slight incline for easy runoff. The idea was that any excess water in the soil would percolate through holes in the top of the drainage pipes and thereby feed into collectors and drainage ditches that emptied into the nearest river, in Buzet's case, the Tarn. It was a way of sopping up surplus water in a field as quickly as possible in order to facilitate effective sowing, favorable germination, and plant growth and thereby ensure good harvests. The pipes had once been baked clay; now they were mass-produced plastic. The trenches to bury them had once been dug with pick and shovel; now the mechanical trenching equipment used electronic and laser guidance systems that worked to a tolerance of 1 centimeter. One of the oldest agricultural improvement techniques had gone high-tech, yet the goal had not changed in two thousand years: to dry out soggy soil.

Though the dynamic young farmers of Buzet may have been convinced that drainage was a good thing, they were, as we noted, less than convinced that they could afford to take on yet another long-term financial burden. It was Maquille, the extension agent, who stumbled upon a solution. By a neat twist, it was back to the European Economic Community! Some time in 1978 or 1979, Maquille happened to come across a 1976 circular from FEOGA (Fond européen d'orientation et de garantie agricoles, the European Fund for Orientation and Support, an agency of the EEC) that outlined EEC-subsidized "development plans." The idea was to encourage European farmers to submit a five-year plan of rational investments designed to bring up their income to at least the average yearly income of all economically active persons in their particular region. The EEC, in turn, promised subsidized low-interest loans to finance this five-year investment plan. Yet what made the offer even more attractive was the promise that, if a certain critical mass of farmers in a given community all filed such plans and included in their individual plans a collective improvement scheme as well, FEOGA would not only provide cheap credit to individuals, but also shoulder 20 percent of the normal subsidies granted to such projects by the national government concerned *and* advance 15 percent of the total cost as a grant to get things started.[45] Maquille convinced the Buzet farmers that it was worth an application, yet he received no encouragement from the officials of the French Ministry of Agriculture in Toulouse, who were slaves to deductive logic: since such a thing *had* never been done in France, this obviously proved that it *could* not be done. Shortly afterwards, Maquille shared a train ride to Paris with a Parisian agricultural official of his acquaintance, who agreed that Buzet ought indeed to qualify under EEC guidelines.

The persistence of Maquille and the Buzet farmers finally enabled them to enlist the help of the DDA, as well as that of an association recently founded to promote the filing of such development plans. Each Buzet farmer's needs and possibilities were analyzed by computer simulation to indicate just what investments undertaken in what time span and in what order would yield the best results. Each farmer was presented with the plan designed to optimally increase his income over five years. The hydraulic service of the DDA agreed to forward the collective application for the twenty-seven Buzetois farmers who had not only filed development plans, but who had also joined the newly formed Buzet Drainage Association.

Yet once again the Buzet Drainage Association seemed to be reenacting the tragicomedy of the Irrigation Association's quest for EEC manna. At this point, however, coincidence entered the game: Maquille learned one day that the very EEC bureaucrat who had signed the auspicious 1976 FEOGA circular—the official in question happened to be French—was coming to Toulouse to give a speech. Fifteen angry Buzet farmers, accompanied by Maquille, greeted him, demanding to know why their request was being ignored by the EEC. The beleaguered and embarrassed technocrat tried evasion and was publicly and humiliatingly attacked by the Buzet farmers as betraying French agriculture, lying through his teeth, and failing to earn his keep. It made a certain impression. Two weeks later his deputy contacted Maquille. Two months later, the Buzet Drainage Assocation received a check for Fr 1,150,000 as their 15 percent subsidy, a sum that permitted the plans for a collective drainage network to go forward.

The drainage project was carried out over the course of three years, beginning in the winter of 1981–1982. The first results lived up to expectations. Even though some farmers complained that the contractors had to be closely watched to insure that all the pipes listed in the estimate were actually installed, those whose land was drained the first year recognized major changes for the better. The Reluc brothers, who occupied some of the shallowest land adjoining the forest of Buzet, had always had trouble in the fall and winter. They were accustomed to finishing their corn harvest by hand, because sooner or later, corn-pickers brought in for the harvest would get bogged down. Yet once drainage had been installed, they were able to spray nitrogen in January without getting stuck, something that had never happened before. The effect on winter crops was equally notable. Ardasse, the president of the Buzet Irrigation Association, who was generally recognized as one of the best farmers in the *commune*, had never done much better than 3 tons of winter wheat because of excess water throughout the winter months. The year after drainage, he averaged 5 tons, as did Grasouillet, his neighbor. In fact, in one field Grasouillet obtained 7 tons of wheat per hectare, which put wheat, a crop that had been written off as hopelessly unprofitable in Buzet, back into the running. Irrigation enhanced by drainage meant, above all, security. A conscientious farmer could now count on his crop, where before, harvests might vary in a ratio of four to one from one year to the next.

Irrigation, whether individual or collective, also dictated a much more intensive and much more profitable agriculture. This new high-cost, but high-yield, agriculture provided the new generation of Buzet farmers with choices that their fathers had never had: polyculture was no longer needed either as an insurance policy or in order to maximize income. For the first time, local farmers could choose on the basis of personal inclination. The EEC's famous development plans made specialization almost mandatory. There were two obvious choices to be made: one concerned milk, the other concerned wine.

Among the native Buzetois farmers of the new generation, cows were unpopular, even though they had grown up with them. Though imbued as they were with notions of economic rationality and maximum return on their labor and capital, the decision to get rid of milk cows was rarely an economic decision. The local dairying tradition was shallow: it went back a mere twenty-five or thirty years. Dairy farming had been dictated to the older generation by an expediency that no longer applied. As the tractor had made dairy farming almost mandatory, irrigation made dairy farming dispensable. For a few, the investments to be scheduled for the five-year development plans were the ultimate moment of truth: to modernize or not to modernize cowbarns was really a decision to specialize either in dairying or in grain farming. Among native Buzet farmers under the age of fifty, the decision was unanimous. Whatever their rationalizations, they gave up dairy cows because they didn't like them and the confined life that they represented. From here on in, they chose to be grain farmers.

With nearly the same degree of unanimity, the Aveyronnais and the North Italians (mostly of French birth by now) chose dairying. Here too regional traditions, rather than profitability, seem to have made up minds. At the same time, irrigated agriculture was a potential revolution for dairy farmers in that it permitted a type of intensive fodder production allowing an extraordinary increase in the number of cattle per hectare of land. Paradoxically, as the number of Buzet dairy farmers shrank the number of cows in the township rose because the remaining farmers operated on a so much larger scale than previously.

The old system of dairy farming still practiced among the older generation of Buzet farmers relied on pasturing in the summer and dried *luzerne* or alfalfa hay in the winter, supplemented with a bit of barley flour
and—a modern touch—some soybean cakes. With this method the Favière farm of 17 hectares managed to keep a dozen cows. The new system
was based on double-cropping made possible by irrigation and the use of
silage in the place of hay. On irrigated land, these dairy farmers planted
corn that was cut while still green and chopped up for silage. Immediately following, on the same land a cover of Italian ryegrass was put in
and allowed to grow throughout the winter and the first month of spring.
This ryegrass joined the corn as silage, while a new spring crop of corn
was sown, fertilized by vast amounts of cow dung to replenish the humus.
For all practical purposes, cattle were on a permanent diet of silage—the
bovine equivalent of canned food, if you will—supplemented with much
larger doses of flour and soy cakes than they would have received in Favière's barn. In practice, the most modern of the Aveyronnais dairies,
the Bechette farm, managed to keep 160 head of cattle—90 milk cows
and 70 heifers—on 66 hectares of land, three times the cattle/land ratio
of the Favière dairy operation. The difference in milk production was
equally striking: Favière's traditionalist cows average 2,400 liters of milk
per year; the Bechettes' 6,700. The difference was not merely one of
feeding, but also of a high-powered breeding program, complete with
expensive frozen sperm from super-bulls and implanted embryos from
super-cows. Between the increase in cows per hectare and their yield in
milk, the Bechettes' gross productivity was about six times that of traditionalist milk producers like Favière.

Yet all this forcing did exact a certain price. Favière's cows calved year
after year until they were eighteen or twenty years old. The Bechettes
were lucky if, with a veterinarian dancing in continual attendance, theirs
lasted three or four calving seasons. The Favières, though they lived
pinched lives, prided themselves on paying their own way; the Bechettes
lived very expansive lives on top of a mountain of credit: their new dairy
installation alone had set them back Fr 500,000, even though they had
done all of the work of construction themselves.

The second decision, to abandon wine, only seemed less clear-cut because it was made in installments rather than all at once. We saw earlier
that by 1913, vineyards in Buzet had more or less stabilized at 170 hectares, one-third of what they had been immediately before the phylloxera
blight. There were some fluctuations over the next forty years, but there
were still 160 hectares left in 1954–1955. We examined how different
events and trends converged to cut that acreage in half by the end of the
land consolidation process in 1970.[46] Only nine years later, in 1979, the
new agricultural census recorded another drop of almost 50 percent to 45

hectares. By the winter of 1982–1983, it was estimated that only 20 or 25 hectares of grapevines survived. At that time the consensus was that within two or three years, a few plots for family consumption and a few hectares of grapes on nonirrigated farms would be the only remaining memorial to Buzet's wine industry.

No one cause fully accounts for the rapid disappearance of Buzet's vineyards in the 1980s. The continuing price decline for ordinary wines when inflation is taken into account was the most obvious long-term reason why Buzet farmers abandoned wine. In some ways, the price situation for Buzet vintners was unusually frustrating. Take the situation of Ardasse who had heeded the latest wave of official advice and had planted noble, but sparsely bearing, varieties of grapes—Cabernet Sauvignon and Syrah—a dozen years earlier in the hope of selling his wine as *vin de pays* (wine of local character) that sold at several notches above the ubiquitous *vin de qualité courante*, a fancy label for the lowest-priced bulk wine. The hitch was that *vin de pays* had to attain a minimum of 9.5 percent alcohol or else it was rated as ordinary wine. For twelve out of thirteen years, Ardasse's wine remained stuck at 9.4 percent. The great experiment of growing noble wines in Buzet merely led him to sell smaller quantities of better tasting wine at the usual miserable price. There was nothing new in that situation. What *was* new was that suddenly it had become immediately profitable to pull out grapevines, because the EEC, reacting to Common Market overproduction, was offering subsidies for abandoning wine growing altogether. These were very generous, up to Fr 30,000 per hectare, about the going land prices in Buzet or the equivalent of two and one-half crops of irrigated corn. There is no doubt that the timing and the rapid pace at which vineyards were being turned into fields owed a good deal to these subsidies.

In the long run, the vineyards of Buzet were doomed, subsidies or no subsidies. Not only did high-intensity, irrigated farming make subsidiary agricultural activities less essential, but the acreage that most vineyards occupied was the land most suitable for irrigating, because it was well-drained gravel. Recassé, for example, knew very well that if he pulled out his remaining ten-year-old vines that winter, he could count on 85 bushels of corn per hectare on the same plot of land. Vineyards also competed with the sprinkler systems for the farmers' time. Cadoulenc, for example, who had no help, quit not because he could not trim his vines— a winter task during a time of the year when other work was slack—but because he could not find the time to spray his grapes during the summer months. By departing from the required rigid spraying schedule, he had lost half his crop and, with it, whatever financial incentive remained to growing grapes. For others, vineyards were abandoned when an old father or a hired retiree stopped working. Curiously, in at least one case the clinching argument for quitting was the coming of new technology, specifically, the introduction of mechanical grape-harvesters. For many small vintners only mutual aid in harvesting each other's grapes had made wine even a half-way profitable proposition. The coming of the mechanical grape harvesters disrupted that tradition of mutual aid, which suddenly appeared as something old-fashioned, or even retrograde. Nineteen eighty-two had been the first year that Recassé had had no choice but to hire a contractor's mechanical grape-harvester. The machine had cost him Fr 2,400. In the past he had been able to rely on neighbors and would have spent no more than Fr 500 to 600 on getting his crop in.

Though grain farmers and dairy farmers took different routes, they continued to meet in the Buzet Irrigation Association and the overlapping, but distinct, Drainage Association. The dairy farmers stayed in touch by helping each other twice a year in putting in silage, an operation that was most efficiently done with five or six men and four tractors. Yet they resisted formalizing these arrangements: Maquille, the extension agent, was unable to persuade them to institute a "labor bank" complete with accounting of work rendered and owed at the end of each year. In contrast to the old conviviality of the peripatetic threshing machine, this mutual aid excluded women. This was also true of the grain farmers, who kept up some of the "group farming" traditions of the Cercle des Jeunes Agriculteurs. On an everyday level, the casual exchange of agricultural information and advice on the telephone was very common, a practice unthinkable in the other townships I came to know.

The most visible recent results of group farming were the immense corncribs scattered throughout the Buzet countryside. The grain farmers had calculated that they laid out considerable sums of money each year to have their corn kernels oven-dried by the coop, when natural drying by wind and weather could be shown to cost much less. Having come to this joint decision, all the grain producers inscribed corncribs in their individual plans of development. They then designed a corncrib made up of identical modules that was durable as well as easy and inexpensive to fabricate. Together they bought materials and tools and together they erected all of the individual corncribs, sized according to each participant's acreage. Since crib drying presumed harvesting corncobs, which called for corn-pickers as well as corn-shelling machines, rather than the harvester-combines used for bringing in corn as grain, the needed new equipment was also cooperatively purchased and utilized. Even the annual emptying of the cribs and the shelling of the cobs was a joint effort. The only thing that the grain farmers had neither anticipated nor calculated was the wave of envy these naked displays of agricultural wealth unleashed.

In their own view of themselves, the farmers who had taken over from their elders during or shortly after land consolidation had always tried to do the right thing. They had chosen farming and had trained to become proficient professionals at it. They worked hard, collaborated with each other, reasoned out their problems, planned, and invested. They were, as the local agricultural technicians never tired of pointing out, model farmers who had taken a mediocre agricultural setting, Buzet-sur-Tarn, and transformed it into one of the most dynamic and prosperous agricultural communities of the region, a showcase *commune* that had become blasé about the delegations of admiring official visitors that streamed in and out. Everyone had been invited to participate in irrigation and drainage; no one had been excluded from the enterprise. As modern farmers, they had relied on brains, energy, persistence, and shrewdness. They had injured no one, asked nothing of the municipality and had always contributed the bulk of the local taxes. If they had persuaded the *commune* of Buzet to restrict random building by passing a zoning ordinance, they had merely protected their capital, the land, from spoliation. And it was only fair that after all these years of sustained effort, they should reap their reward with incomes on a par with middle-level executives'. Unlike such executives' salaries, however, year after year, one-half and sometimes even two-thirds of their income was being reinvested in improvements and equipment. At the height of the agricultural season, moreover, they might put in *their* thirty-nine hours—France's legal work-week in 1983—between Monday and Wednesday and put in another thirty-nine hours before the week was over. And if they dried their corn in open cribs for all to see, it was certainly not to show off their affluence, but to take advantage of wind and sun as being cheaper than gas-heated drying sheds.

Yet this was not how everyone saw them. Hostility did not come from the remaining marginal farmers, who, with a few exceptions had been too old to launch into daring bootstrap operations and who were waiting for or already collecting their old-age pension. There was, to be sure, irritation from the commuters, whose notion of bucolic peace—*le calme* so prized by those who are rarely calm themselves—had not reckoned with the unmuffled engine roar from 6:00 A.M. to 10:00 P.M. as normal auditory background to rustic living. The real drama was the seething hostility of the retirees living in the village. This was a subterranean class struggle of the ex-farmers, those who had abandoned the unequal contest with a meager soil twenty or thirty years earlier, only to return "home" as pensioners, against the farmers who had stayed on and made it, visibly, conspicuously, with heaping corncribs as year-round testimony.

And, to add insult to injury, some of the younger generation had become this successful by leasing the very land on which those who were now pensioners had failed long ago. And the members of the Buzet Irrigation Association, who acted as though they owned Buzet, had made sure with their zoning that the pensioners would not even be able to sell the field or two they retained as building lots. Understandably, the latter found the situation galling.

The municipal elections of March 1983 brought the accumulated bitterness into the open. There were vicious handbills and even nastier graffiti: "Down with the landgrabbers! No to the peasants! We've had enough of the farmers! Nothing but agriculture counts in Buzet! They've given the farmers too much!" The first election results caused genuine dismay in incumbent ranks. Among the members of the incumbent Unity slate, farmers had been systematically singled out: they averaged 25 percent fewer votes than nonfarmers. At the runoff the following Sunday, five opposition candidates from a slate of fifteen with not a farmer among them eliminated five of the seven farmers who had sat on the previous municipal council. Only two farmers survived as municipal councilors, one of whom was a nonirrigating farmer whose modest life-style challenged no one.

To Mallorca, the problem was simply one of poor public relations, of being too busy to fraternize, too busy to come to the café, too busy to explain the problems they faced. The losers themselves were genuinely hurt and bewildered to discover that their success had brought them hostility and envy, rather than the appreciation and respect that they had felt was their due. For the first time, they felt beleaguered on their own turf.

The experience awakened half-admitted fears that they were being swamped, that zoning had come too late, that too many outsiders had already built within the township, that Toulouse loomed too large, that encroachment was inevitable. Cadoulenc, a buoyant personality if ever there was one, spoke gloomily about his fears of being unable to preserve his land for his children. Perhaps they had all been fools after all. Perhaps they still ought to sell out to the commuters, make a killing in real estate, and then start out all over again as farmers somewhere else, so far out in the countryside that the Toulousains could not follow them this time. The municipal elections had had a sobering effect.

2

Requiem for a Mountain Village

EVEN though the name of Juzet rhymes with Buzet, the resemblance between the two communities begins and ends there. Juzet d'Izaut is a village in what the French call *la pré-montagne*, the fore-mountains of the Pyrenees, located at an altitude of over 1,900 feet.[1] It nestles at the foot of a 6,000 foot cone-shaped peak called the Cagire, that looks as though it had been drawn by an unimaginative six-year-old. The village is traversed by departmental route 618, the last French east-west road paralleling the Spanish border, a boundary line that follows the crest of the high Pyrenees that lies 20 to 30 kilometers further south. This very scenic mountain road, which ultimately snakes its way over the Ares pass, is well traveled during the summer tourist season, but desolate the rest of the year. Juzet's one restaurant, a recent arrival that may or may not survive, therefore opens only during the summer months, as does the communal *gîte rural*, the village-owned tourist home. The village has a bakery, as well as two wholesale *charcutiers*, who produce the usual line of French pork products from sausages to *pâtés*, but whose business turns out to be unrelated to local agriculture. M. Hughet, the owner of the larger of the two, employing four or five people, has been mayor for the last twenty years. The grade school continues to operate, though one wonders for how long.

The 1977 official figures claimed 248 inhabitants, but that is surely wishful thinking: two-thirds of the houses have become vacation homes that stand empty all winter (the mountains being too wooded for winter sports). Their owners' claim to be Juzet d'Izaut residents bloats the local census, but has otherwise only a tenuous impact on the village. There are probably less than 100 year-round inhabitants and many of those are pensioners. Juzet gives the forlorn impression of being out in nowhere, yet Saint Gaudens, a bustling, industrialized city that also houses the subprefecture, is only 15 miles away and even Toulouse is less than 70 miles away.

Just how many farmers are left in the village is also a matter of some controversy. The official Chamber of Agriculture statistics count as many as ten, but they include Mme Peyras who was born in 1902 and M. Bensaque, who is physically handicapped. Mme Dessault, who is also featured on the official list, grazes horses on land that she inherited from her parents, but she really lives over in Izaut l'Hotel, the village a couple of miles north of Juzet, where she is postmistress and runs the grocery store with her husband. The statistics also include a family like the Arrêtes, who raise cattle that pasture out of doors all year. In reality, the only members of the family living on their farm in Juzet are the grandparents, who are octogenarians. Their middle-aged son, the man who actually takes care of the steer, merely drops by two or three times a week. He and his wife are grade school teachers in a community 15 kilometers away, though he has just built himself a villa in his native village in anticipation of retiring there at age fifty-five. And there are other part-timers: postmen, road commission workers, forest service employees. Purists insist that there are only two or three genuine farmers left, if by that you mean families who receive all of their income from agriculture. And one and only one of these has a son likely to take over the family farm.

Traditionally this had been an agricultural community and one that, by mountain standards, has some natural advantages. Although, as elsewhere in the area, the growing season is short, drought is hardly ever a problem. Unlike many other villages in the Pyrenees that cling precariously to their steep slopes, a good part of the *commune* varies from gentle incline to relatively level plateau, some 460 hectares in all, jammed between the Cagire to the south and a 1,000-foot high hill usually called only *le mont* to the north. Since World War I, the forest has been allowed to gradually encroach upon this farming oasis, which, within the memory of old peasants, has yielded 50 to 200 yards along its edges. As older farmers like to point out, as late as thirty or forty years ago, every July you could still see a wide swath of golden wheat stretching for almost two miles. Although much of this land was considered suitable for grain, corn, and potatoes, almost all of it is now a lush green three seasons of the year. Pasture and natural meadows have displaced the fields.

Unlike Buzet with its complicated social structure manipulated by wealthy outsiders, eighteenth-century Juzet d'Izaut was much more primitive. The king in the person of the royal judge in Aspet, 4 miles away, was lord of Juzet, but the only outpost belonging to outsiders was a manor house with its dependencies that until about 1760 belonged to a noble, Gabriel de la Motte d'Izaut, who passed on his chateau and the rest of his property—about 15 percent of the *commune*'s usable land, including its best meadows—to the monastic chapter of Saint Lizier.[2] A decade or two later there was a second transfer, this time to the Seminary of Saint Gaudens that was the only significant absentee landlord at the time of the French Revolution.[3]

Aside from this one powerful outsider, the population consisted only

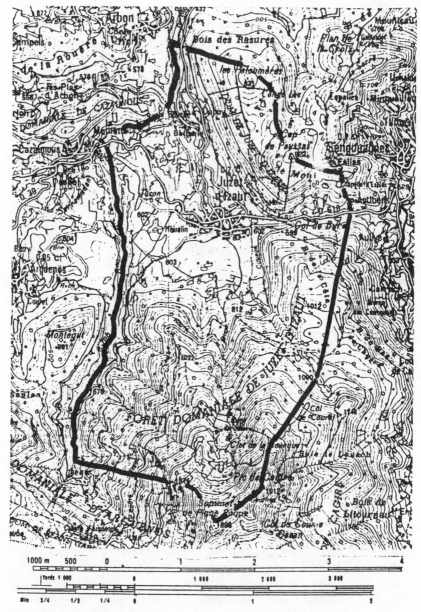

8. The township of Juzet d'Izaut

of peasants at varying levels of poverty, reflecting the terrible pressure of a dense population on too little land. A whole series of tax assessment rolls from 1760 into the 1780s all tell the same story. Except for the 55 hectares held by the seminary, the land was unevenly distributed among well over one hundred peasant families—117 of them in 1782—although there may also have been propertyless paupers who did not show up on the village tax rolls. What is striking is not the maldistribution of property, but rather the fact that even the "richest" peasants were in such marginal circumstances. In Juzet the two wealthiest peasant families, assessed at a yearly income of just over 100 *livres* (to give an idea of the purchasing power of the regional *livre*, according to a contemporary tax official, local day laborers earned 1 or 1.25 *livres* per day during harvesting), owned farms of 6.5 hectares (16 Anglo-American acres), farms that were very close to minimum subsistence. The *median* land holding, moreover, was less than 1.5 hectares—perhaps one-fourth the amount of land it took for family survival—in a community where, with the single exception of the one collective absentee landlord, even the wealthiest peasants had neither the need nor the means to hire those even more impoverished than themselves.[4] In 1761, the royal tax collector made a point of noting that no trade was being carried on in the village, which may explain why the same tax collector, as well as his successor twenty years later, sympathized with these peasants whose landed income did not stretch to pay their taxes. And this was true even though, aside from meadowland, all but the very worst soils were planted to crops every year without fallowing.[5]

How desperately poor the peasants of Juzet were shows up in the virtual absence of artisans in the village: they simply did not have the wherewithal to pay for specialists. The only nonpeasants listed on the tax roll of 1789, for example, were a tailor, a surgeon, and a musketeer, presumably a discharged veteran.[6] A village of that size that could not afford a blacksmith is astounding, but apparently iron-reinforced plows were not introduced until the nineteenth century, horses were few, and draft-oxen and cows must have gone unshod.

How then did the 600 or 650 inhabitants of Juzet manage to survive at all in the second half of the eighteenth century? It was certainly not because their grain fields produced very much. Indeed, even Buzet's low early–nineteenth century yields seem lush by comparison. In 1782, the tax collector figured that the seed/crop ratio for Juzet's first-class land was 1 to 3.75, with a net production of about 450 kilograms of wheat per hectare. Class II and III land was considerably less productive. What Juzet peasants could and did do was to find ways of stretching animal feed. All summer long, sheep and those oxen and cows not needed for carting grazed beyond the tree line on the alpine pastures of the Cagire,

guarded by the communal herdsman. This meant that the valley meadows could be devoted to making hay, rather than pasturing cattle, which,
in turn, meant that more head of cattle could be sustained through the
winters. Hay fields were further extended by clearing the lower reaches
of the forest, even though the incline of these meadows was too steep for
any cart or sled and peasants had to bring some of their hay to their barns
in huge bundles piled on their heads. Even though, then as now, the vast
forest of the Cagire—over 1,000 hectares lay within the boundaries of the
community—belonged to the state, the forest was nonetheless a major
asset. Since time immemorial, the residents of Juzet had had access to
the forest to meet their needs for wood for heating their homes and for
building lumber. Since only mule tracks connected Juzet d'Izaut with
market villages like Aspet, not to mention the town of Saint Gaudens,
even had they owned the forest, it is doubtful that local timber was anything that could be marketed.[7] Yet in any event, traditional forest rights
did mean that the poorest of the poor could count on free fuel and that
everyone had access to the lumber needed for tools, equipment, and
shelter.

Access to additional pasture, to firewood, and to construction lumber
would not have been enough to support Juzet's dense population. The
main safety valve appears to have been seasonal migration to Spain.
Those too land poor to make it by agriculture, the eighteenth-century tax
collectors reported, spent five or six months of the summer working in
Spain.[8] During the nineteenth and early twentieth century a different
pattern developed, as each year a sizable portion of the male population
now headed north rather than south, working as itinerant tinkers—that
is, repairing tin pots and pans—or as scissors- and knife-grinders.

What we know of the evolution of Juzet in the first half of the nineteenth
century is spotty and indirect, but two kinds of information are sugges-
tive. In the first place, the already excessive population continued to increase from 716 in 1809 to a peak of 957 in 1836, only to begin its long
decline that may not yet have ended.[9] In the second place, by the late
1820s when the first postrevolutionary land register was drawn up, the
pattern of landownership had changed markedly. Though we lack information, evidently most, if not all, of the 55 hectares of local church lands
sold during the French Revolution must have ended up in peasant hands.
Aside from a few peasants from adjoining villages, there were no outsiders owning landed property in Juzet in 1828, the year the land register
was completed. The chateau of Lamoulette, formerly owned by the Seminary of Saint Gaudens, had been allowed to fall into ruins.

A more striking contrast to conditions in the eighteenth century was
the widening of the gap between the top and the bottom layers of the

peasantry. We can only speculate as to causes. Did the sale of church land split peasant society in Juzet? Did improved access to markets in Aspet and Saint Gaudens enhance the position of those with even a minor marketable product? Did increasing population pressure also help widen social distances? Whatever the reasons, the facts themselves are incontestable. Compared to the 1780s, when only two peasants owned as much as 6.5 hectares, by 1828 there were all of seven proprietors—six peasants and one grain dealer (whose presence was itself indicative of change)— with farms between 10 and 17 hectares, that is, by nineteenth-century standards, large and presumably prosperous farms. Below this top crust were another fourteen owners with comfortable farms between 5 and 10 hectares. These twenty-one families controlled almost one-third of Juzet's land. In the middle were 108 owners of land between 1 and 5 hectares, whose total holdings amounted to 57 percent of the *commune*'s acreage. Finally, at the bottom were 112 proprietors of less than 1 hectare, whose share of the land was a mere 12 percent.[10] Inequality had far more bite in 1828 than it had had fifty years earlier.

Poverty and overcrowding were not peculiar to Juzet d'Izaut, but characteristic of the entire canton of Aspet. The underemployment of the agricultural population was a chronic ill, with seasonal emigration the norm. Wages had sunk lower than they had been in the eighteenth century—a man received Fr 1 a day in 1848, a woman 70 centimes—but the local cost of living was also low. Peasants with land could manage with a supplement of only 50 centimes in cash per day, while a family with two young children and no land could, it was claimed, get by on a daily Fr 1.25. One common complaint was that communications with Spain were so poor that local farmers were cut off from their best potential markets.[11]

During the second half of the nineteenth century, population pressure began to diminish in Juzet. This was partly due to the fact that for the forty years between 1845 and 1885, local births and deaths were almost evenly balanced, with 592 births to 604 deaths, thirty-two of which may be accounted for by a cholera epidemic in 1855. During the 1850s and 1860s widespread, and, more often than not, permanent, emigration to America by young men led to a sharp decline: between 1845 and 1872, Juzet went from 923 to 776 inhabitants and slid further to 715 by 1881.[12]

This emigration was distinct from the continued seasonal migration that had, as already noted, shifted direction since the eighteenth century. At the beginning of March, bands of men, but including boys as young as ten years old, set out on their yearly trek to the Limousin, some 250 miles north of the Pyrenees, to pursue their trade as itinerant tinkers, until they returned before the onset of winter. As late as 1896, about one male out of five in the fifteen to fifty age group was a seasonal migrant.[13] Just why they chose that particular destination, the documents do not say. Ironically, the Limousin was another impoverished and overpopulated region: *its* surplus males went to work as stone-masons in Paris and Lyon. Apparently, for a time, some of the Juzet migrants virtually ceased to be peasants. Instead of leaving the summer farm work to their wives and children, they hired sharecroppers from the neighboring *département* of Ariège to run their family farms for them. However, this practice, which the local schoolmaster deplored as unnatural, had lost favor by the mid-1880s.[14]

Agriculture had changed for the better since the eighteenth century. Reinforcing the traditional *araire* plow with iron had become common practice by midcentury, while crop yields had distinctly improved, perhaps because more cattle made more manure available. By 1884, the net yield for wheat, for example, seems to have doubled from what it had been a hundred years earlier: it went from 450 to 920 kilograms per hectare. More important, wheat was no longer the dominant crop, having been partly supplanted by the much more productive potato. Under the pressure of population, Juzet had taken the same route as nineteenth-century Ireland and southwestern Germany. Although half the arable land was still planted to wheat and only a quarter to potatoes (with the remainder in corn), Juzet d'Izaut harvested 13,000 sacks of potatoes as compared to 1,700 sacks of wheat and 1,300 of corn. Wheat, corn, and potatoes constituted the normal crop rotation in the township.[15]

Compared to undoubted improvement in the local crops, other agricultural changes were less clear-cut. From 1809 to 1872, the number of cattle almost doubled to 400, but then declined to 315 by 1885. Given the fact that only 260 hectares were under the plow in Juzet, there were too many head of cattle for all of them to have been draft animals. By 1885, the local schoolmaster made a mention of occasional calves and a few pounds of butter for sale at the village market (itself an innovation, as were the two small groceries and the four taverns that only filled up on Sundays). In any event, cattle now provided enough dung so that fields were regularly manured with 1,800 kilograms of manure per hectare. Sheep could only be kept as long as they were allowed to graze in the mountain pastures. Their numbers fell from 500 to none within a few years after the State Forest Administration, temporarily as it turned out, outlawed sheep grazing on the Cagire in 1883.[16]

As was true in many parts of the Pyrenees, forest rights became the major local grievance.[17] Juzet's original charter of forest rights apparently went back to 1335 and these rights were confirmed twice in the seventeenth century and once by a Napoleonic prefect as late as 1807. They came under attack at the end of the 1820s as the controversial and harshly restrictive new forest legislation went into effect.[18] In the 1880s local people were still complaining bitterly that, despite their historic and officially recognized right to free firewood, they were only allowed to cut five cords per family, when the average family needed twenty cords per year for cooking and heating.[19]

On the eve of World War I, at a time when the population was about 450—less than half of the inhabitants that had lived in Juzet two generations earlier—the acreage of crops harvested in the *commune* was virtually unchanged from what it had been in the 1880s: 120 hectares of wheat, 63 of corn, and 61 of potatoes, which meant that the far more productive potatoes weighed heaviest in the local food balance. There was even more land in meadow and pasture: 206 hectares of natural meadow, 9 of artificial fodder crops like clover and *luzerne*, as well as 150 hectares in pasture, not counting the summer pastures on the Cagire. Even though the number of cattle had declined slightly, from 315 in 1885 to 273 in 1913, because the population had declined even more sharply— from 700 to 450—on a per capita basis peasants were somewhat better off. There is also evidence that dairy production was being taken seriously: two local dairies producing butter were listed in the agricultural census. The 1913 statistics do suggest lessening poverty for those who were staying on: most families, for example, seem to have fattened one or two yearly hogs.[20] Yet conditions were still primitive in ways no longer found in the plain: peasant families, for instance, had to bake their own bread.

In some ways the period between the two world wars, the 1920s in particular, may have been a modest sort of golden age for Juzet's inhabitants, though it is undoubtedly illuminated by the nostalgic glow of old people's recollections. Yet even if we discount some of the more obvious nostalgia, life was indeed becoming easier at a time when rural expectations did not run very high. On the humblest level, between the wars the self-adjusting Brabant plow replaced the iron-reinforced *araire*, which, earlier, had superseded the wooden plow of the eighteenth century. Two-thirds of the peasant households had acquired simple harvesters for cutting grass and grain, though mechanical rakes and haying machines were still a rarity in the 1920s.[21] Substantial farmers could now afford the more powerful work-oxen to supplement the labor of cows: the man of the household would lead a pair of oxen pulling one plow, while his wife would plow with a team of cows. Everyone kept cattle: three or four cows were common, five or six average, seven or eight the maximum. Peasants grew or raised almost all they consumed: everyone slaughtered one or two yearly pigs, inviting the neighbors to the feast, just as they invited neighbors to winter corn-husking bees at which chestnuts were roasted and young people danced after the work was done. Ducks and geese were fattened for family consumption, as they could be easily preserved in their own grease as *confit*, just as hog meat kept as salted pork and sausage. Calves and lambs, on the contrary, were invariably sold and never butchered for home consumption. Yet in the early 1920s, at the wedding festivities of one of the village's substantial farmers (heir to a farm of less than ten hectares), six musicians were hired to play for two whole days, something that farmers with three or four times as much land would not dream of affording in the 1980s.

Some of the amenities of modern farm life were introduced in the 1920s: this was the time when, for example, electricity was brought to Juzet d'Izaut; by 1929 the village's two dairies, one specializing in butter, the other in milk, had such aids as electric mixers and cream separators. The prewar threshing machine, powered by a water mill, to which peasants had had to carry their sheaves of grain, was replaced by the typical steam-thresher used in the plain that was hauled from farmyard to farmyard, thereby saving peasants the work of carting the sheaves. In turn, the steam-powered threshing machinery gave way to less cumbersome equipment for which a tractor provided the power. Bus connections to Saint Gaudens made going to market easier, though cows still had to be taken on foot, a round-trip that began at 3:00 A.M. and ended at 11:00 P.M. For a decade or two, this relative prosperity permitted Juzet to have its full complement of village artisans—three blacksmiths, various building tradesmen, a baker, a butcher, a clogmaker, a shoemaker, as well as several shopkeepers.[22] Rather than continuing to have to bake bread at home, peasants now took their grain to the baker, swapping wheat against bread pound for pound.

Paradoxically, this prosperity was directly related to the rural exodus, begun in the 1840s, that had reduced population by somewhere around 10 percent each and every decade prior to the war. It accelerated between World War I and the 1930s, shrinking the village's population by 40 percent in a little over twenty years. By 1926, Juzet d'Izaut was down to 365 inhabitants; by 1936 to 277.

Yet in spite of, or perhaps because of, its sparser population, Juzet no longer appeared so completely dependent on combining subsistence farming with migrants' earnings: during the 1920s one-third of the inhabitants engaged in local economic activities other than farming. About two-fifths of these nonpeasants relied on work connected with lumbering and the forest. Itinerant tinkers and scissors-grinders went the way of other folklore, but in 1926 two residents were described as "traveling salesmen" by the census takers.

The presence of nonagricultural workers turned out to be a transitory phenomenon, as the interwar rural exodus resulted in the depletion of their numbers: most of the nonagricultural workers, artisans and lumberjacks included, were gone by 1936, whereas the number of farm households was down only slightly, from sixty-nine in 1926 to sixty-one in 1936.

Half of these owned farms of more than 4 hectares, farms that comprised 42 percent of the *commune*'s land. Because so many of Juzet's 357 proprietors had left for good, some of the remaining farmers may have been able to lease additional land to round out their holdings.[23] In that sense, emigration had helped appease an age-old land hunger in Juzet.

Yet emigration also changed the agricultural landscape. As some emigrants' ties to "their" village became increasingly attenuated—after all, some had left generations earlier—scattered fields and meadows, neglected because their owners' whereabouts were unknown, became a conspicuous nuisance for Juzet's peasant farmers.

In turn, the decline in the agricultural population affected land use. In 1913, 257 hectares had been under the plow; in 1921 that figure had dropped to 146 hectares, the level at which it stabilized during the interwar period. Among crops still grown, wheat had fallen off most drastically to two-fifths of the 1913 acreage; corn and potatoes had declined less sharply to two-thirds of the land devoted to these crops prior to the World War. Most of these harvests continued to feed the local population, yet some were marketed: one old Juzet farmer recalled days when 10 tons of potatoes would be carted to the Saint Gaudens market. Some sort of equilibrium between subsistence and market agriculture had been reached, paralleled by a dual shift in land use: on one hand, more land was allocated to meadows and pasture to feed a growing herd of dairy cows—in 1932 there were 380 head of cattle, of which 320 were milk cows.[24] Never before or after was there so much land in artificial meadows, in clover, alfalfa, and *luzerne*, though milk cows were also fed on winter rye mowed green in the spring and mixed with straw.

On the other hand, considerable acreage was abandoned outright, either because there was less need to rely on marginal, hard-to-work land or because, as mentioned earlier, its absentee owners no longer bothered seeing to its cultivation. The taboo against touching another man's property insured that working farmers would leave such land alone. Hence these islets of untended land remained a characteristic feature of Juzet's agricultural landscape.

The interwar equilibrium at Juzet d'Izaut was thus based upon a smaller agricultural population with larger farms that produced almost all of its own food, a homegrown food supply supplemented by cash income from the sale of milk, meat, and potatoes. It was a way of making do, of getting along by carefully husbanding resources—except, as we saw, on special occasions like a wedding. Compared to the naked poverty of the eighteenth and nineteenth century, it was a life of modest comfort, but save for the introduction of electricity, those standards of comfort were pre–twentieth century, based on minimal amenities, minimal productive outlays, minimal household expenditures and uncounted amounts of hard work by all adult and adolesent members of each farm family.

Such an agricultural system could not survive in the environment of the expansive, modernizing economy that emerged in France by the mid-1950s. Attempts to transform this mountain agriculture into something that would be viable in the dynamic new society proved extraordinarily difficult, even though the farmers of Juzet showed more initiative than those in many other mountain communities. By the mid-1970s it was becoming evident that agriculture as a way of life was doomed in Juzet, as it was in many other Pyrenean villages. A transition to agriculture—or rather to cattle and horse raising—as a sideline, as a form of moonlighting, was well under way. This outcome may not have been inevitable, but neither was it fortuitous.

In the first place, personal expectations among this rural population changed dramatically in the decades after World War II. To own a car—unheard of in the long intermission between world wars—came to be taken as a human right. By 1948, the Arrêtes, then as now, one of the two or three most substantial landowners in Juzet, had their *Deux-Chevaux*, Citroën's postwar answer to the American Tin Lizzy of the 1920s. Since the 1960s ownership of an automobile has become a matter of course. When another cattle raiser, M. Ruau, wanted to prove how far he had fallen behind common French living standards, he complained about his inability to regularly trade in his old car, in contrast to what he saw as the average Frenchman, who, M. Ruau was convinced, did so every year or two. By 1960, water was piped into the village, revolutionizing farmhouse interiors with kitchen sinks, hot water heaters, bathrooms, and indoor plumbing. As one farm wife pointed out: "Now we have the water-bill to pay, but who would want to go back to getting the water from the well?" None of the farmers I visited any longer used the traditional fireplace for cooking: kitchen stoves became the norm, as did refrigerators. With a delay of only a decade or two, these ex–subsistence farmers did indeed enter modern consumer society. Finally, eager for its benefits, but reluctant to pay for them, by the 1960s farmers were fully integrated into the highly pluralistic French welfare state. Mandatory payments for the new amenities of health insurance and retirement pensions were also part of the new consumerism. The fact that these payments—referred to as *les charges*, about which all French farmers complain routinely—tended to be regressive, that is, a heavier burden for those on low incomes, made them particularly onerous. Back in "the good old days," Juzet's peasants would never have been able to raise the equivalent of Fr 6,000 per year, the approximate sum of these social security payments for 1982. Even today, in Juzet these obligatory health and retirement contributions absorb 20 percent of the farm income of the relatively prosperous and an even larger share of more modest incomes. These transformations of the conditions of daily life, burdensome in their impact, but welcome in their results, were bound to unbalance what was a delicate equilibrium. The old agriculture simply could not sustain contemporary demands and expectations.

As we already saw at Buzet-sur-Tarn, motorization there tended to push farmers from predominantly subsistence farming to a market-geared agriculture, from low-investment/low-yield, to high-investment/high-yield agriculture. The tractor with all its corollary equipment also called for much larger farm acreage to amortize costly new machinery.

That pattern is not quite so clear at Juzet d'Izaut. In the first place, many farms in Juzet started out at a size well below the optimum even when plowing was done with oxen or cows. Enlargement began because the continuing rural exodus made additional land available for those who remained. The gradual decline in tillage and the increasing concentration on stock raising and dairying also permitted greater expansion for farmers who plowed less and less and therefore were not subject to the chief constraint limiting the acreage of a family farm.

Secondly, modernization in Juzet d'Izaut also tended to precede motorization, since the most striking obstacle to greater income was the low quality of the indigenous dairy cows, rather than the timing or speed of plowing. The local gray cows of the Saint Girons breed were small, tough, and evil-tempered. Above all, they were resilient, which animals that often spent all summer grazing in alpine pastures had to be. Yet they did not average much more than 5 liters of milk a day—1,500 liters a year, less than a fourth of what the most high-powered dairy farmers in Buzet-sur-Tarn produce routinely nowadays—and, when sold as meat, they weighed in at a puny thousand pounds or less. For instance, in order to improve his dairy herd, Arrête senior, who may have been influenced by his experience as a prisoner of war in Germany, bought his first Swiss calf as early as 1948, two decades before he got around to acquiring a tractor.

Thirdly, most tractors not only came a decade later than in Buzet-sur-Tarn—the 1960s was the period of introduction for Juzet d'Izaut, even though as late as 1970, eight farms relied on borrowing or hiring tractors—but they never became so dominant a feature as in the grain-growing plain.[25] By that time, less and less land was being plowed up for crops and no thought was given to increasing that acreage because, unlike oxen, tractors did not have to be harnessed every morning and therefore were less trouble for cutting grass, haying, spreading dung, and, a few years later, spraying fertilizer. Then and later, there was no good reason for powerful and expensive engines. In fact, given the tiny size of the average field or meadow prior to land consolidation, sizable tractors would have been virtually useless. In these circumstances, their introduction was a convenience more than a productive revolution and since the first second-hand tractor was typically an even trade against a pair of oxen, it imposed no special financial burden.

We saw earlier that all farmers faced the problem of adapting to a world in which the terms of trade for the commodities that they produced tended to deteriorate, which meant that they had to keep turning out more and more, even if they merely meant to maintain the real income that they had been receiving. Yet what made modernization so much more problematical in a mountain region was the fact that in such a race, a *commune* like Juzet d'Izaut could not compete on equal terms with agriculture in a more favorable geographical setting. No amount of chemical fertilizer could change the fact that the *commune* was located at an altitude of 1,900 feet, that it had a long, harsh winter, a short growing season, and less sunshine than the plains of the French southwest. Whatever the techniques employed, wheat or corn or potatoes could be produced cheaper, and therefore more profitably, in the lowlands.

As long as crops were primarily intended to meet the peasant family's own food needs, this lack of competitiveness was irrelevant. Even nowadays, the traditional attitudes are reflected in the older Juzet residents' comments on former grain harvests: they will remember that so-and-so used to get wonderful crops of 70 *quintaux* of wheat. If they are asked what so-and-so's wheat *acreage* was at the time, their answer becomes vague, because for them the question misses the point. As long as bread to feed the family came first, you sowed what needed to be sown, rather than worrying about wheat yield per hectare. Yet once a farm family became fully caught up in the money economy, it became a luxury to grow crops that, hectare for hectare, fetched less on the market than, alternatively, the milk or meat that could be produced on the same land. This is why, year by year, Juzet's acreage in wheat, corn, and potatoes declined, as the pattern of subsistence agriculture became less and less viable.

Yet things were not quite that simple. Although it no longer made economic sense to grow wheat for the daily bread in an area where every wheat harvest had been a triumph of the human will, good arguments could still be made for planting and digging one's own potatoes, something that required merely a patch of land and was less expensive than buying at retail. An even stronger case could be made for *some* continued grain and corn growing as an essential complement to modern dairying and stock raising. High-yield dairying was clearly predicated on the use of silage, preferably both grass and corn, a system that, its problems not-

withstanding, permitted both greater milk yields per cow and the maintenance of more cows per hectare. There was also something to be said
for providing one's own barley flour to supplement this diet of silage,
rather than having to buy it at retail. In the case of raising calves or lambs
for sale, having homegrown grain made the difference between selling a
very young animal fed only on cows' or ewes' milk and being able to
market a larger, fattened animal that fetched considerably more. Given
the modest scale of stock raising in Juzet, fattening with purchased grain
was not a paying proposition. Therefore a small acreage of locally grown
corn and grain remained important in helping Juzet farmers to make a
successful transition to modernized livestock raising.

In Juzet d'Izaut it was the clash of hunting with agricultural interests
that increasingly interfered with growing crops. There existed an elaborate system of regional hunting associations, each with its own pack of
hunting dogs, associations that each fall rented—the bidding reached into
the hundreds of thousands of francs—hunting rights in the state forests
surrounding Juzet. Over the years, such associations had made sure that
these forests were thickly stocked with wild boar as well as with the ma-
jestic red stag that they had introduced as more sporting than the little
native roe deer. It would be facile to say that this fight between recreation and agriculture merely mirrored the opposition of city and countryside. The lines were not so neatly drawn. Although Saint Gaudens notables dominated the area's hunting association, they had prudently
coopted the mayor of Juzet as vice-president, while Arrête junior was
proud to serve as keeper of the association's pack of twenty-nine hunting
dogs.

As long as numerous fields of corn, wheat, and potatoes were scattered
throughout the perimeter of the *commune* of Juzet, damage by game also
remained dispersed and manageable. Just as the acreage of cropland contracted in response to changing economic conditions, a growing number
of deer and wild boar concentrated on the shrinking number of fields. By
the end of the 1970s, one farmer complained, he remembered counting
seventeen deer in a single field of corn one evening. That year he had
harvested but a single sack of corncobs from this one hectare field. Beets
and potatoes were increasingly uprooted by boar, wheat trampled.
Legally and theoretically, farmers were entitled to compensation for
game damage from funds furnished by hunters and hunting associations.
In real life, the officials who had to verify such damages rarely bothered
showing up. Besides, as I was told, a farmer's business was to raise crops,
not to collect compensation for their destruction. A vicious cycle set in.
Increasing game damage discouraged farmers from planting; as less was
planted, the more those fields still in crops drew the game of the sur-

rounding forests, which, in turn, discouraged more farmers from putting in crops.

Even embracing cattle, sheep, or horse raising, or else dairy farming, as the economically rational course did not necessarily lead to salvation. Although this might be the best use of mountain land, farmers in the plain could still outperform those in the mountains, particularly when it came to fattening animals for sale or to providing corn silage for milk cows. Where corn and ryegrass silage could maintain two and one-half cows per hectare of land in the plain, even an optimistic agricultural extension agent considered one and one-half cows per hectare the maximum carrying capacity for the canton of Aspet in which Juzet d'Izaut was located. And in real life, local farmers never managed to keep more than one cow per hectare. According to the extension agent's rule of thumb, in the mountains it took forty cows before a dairy farmer made the equivalent of the industrial minimum wage, Fr 3,200 in 1982. Only a single Juzet farmer even remotely approached that number of milk cows. Nor was there any chance of making more money by controlling the processing of local dairy products. The two local artisanal processors, one for milk, the other for butter, that had been in business since before World War I could not survive the competition from the industrial dairy plants in the plain, most of them owned by gigantic regional coops. Both of these farmer-artisans were forced out of their milk-processing business, though the buttermaker switched to fabricating pork products and was successful enough to get out of farming altogether and popular enough to become, for all practical purposes, Juzet's mayor-for-life. Since fattening hogs took grain and less and less grain was being grown locally, the *charcuterie* was no bonanza for the village's farmers: hogs imported from Belgium could be bought cheaper than the local product. Local farmers even gave up raising the traditional yearly family pig.

Making cattle, horse, or sheep raising profitable called for extensive acreage and even larger numbers of animals than milk cows did, as meat provided only three-quarters as much income as milk. No individual Juzet farmer had that much land at his disposal. Moreover, taking the usable land of the *commune* as a whole, what had never been a vast clearing to start with had over the decades shrunk notably in size, as a diminishing farm population fell back in its age-old war against the encroaching forest. In 1913, cropland, meadow, and pasture took in 607 hectares; in the 1960s, when land consolidation was initiated, only 339 hectares were still in use.[26] True, that acreage could be stretched by relying on the summer pastures on top of Mount Cagire, but only at the cost of adding to the farmer's workday, since the village could no longer afford a communal herdsman. Besides, only hardy regional breeds thrived up there and these were not animals noted for either their milk or their meat.

The situation of Juzet's farmers was further complicated because the *commune* had inherited an incredible patchwork of miniature fields, meadows, and pastures from its earlier overcrowded days. As late as the 1960s, at a time when the village's official population had fallen well below 200 and the number of its farmers, full- and part-time, to 21, there were 428 landed proprietors owning 3,268 plots of land inscribed in the cadastre, each plot, picturesquely, being referred to by its given name.[27] It is true that nearly half the land utilized in the nineteenth century had gone back into woods, but even the 339 hectares of land still in agricultural use counted 183 proprietors owning 2,375 plots, which meant that the average field size was a minuscule 0.14 hectares or about one-third of an acre.[28]

This extraordinary dispersal of the land made larger herds difficult to pasture and motorized equipment unusually awkward to employ. On tiny pastures, either herds had to be divided and allowed to graze on different plots, or else farmers were forced to move them continually from pasture to pasture. The smallest plots tended to be cropped for the very reason that under modern conditions they had become useless as pasture, though they might once have provided grazing for one or two sheep. Not even small tractors were designed for plowing fields 30 feet wide, which is part of the explanation for the delay in the introduction of tractors to Juzet. A good many more of the meadows were too narrow for a tractor hauling a trailer to make a U-turn, while the contractors' combines, which everywhere had come to replace the stationary threshing machines by the end of the 1960s, were completely unmaneuverable in some of these miniature fields. Then there were fields that were enclaves within enclaves within enclaves and that were most inaccessible at the very time when access was most needed—when grain should have been harvested, grass cut, or hay raked—because the passage across another field might be obstructed by another crop ripening at a different time. At best access was on paths that oxen had been able to tread, but that barred even small tractors: they were much too narrow and, whenever it had rained, hopelessly mucky. And, as already noted, the area still used for meadows, pastures, and crops was punctuated by abandoned fields owned by former Juzet residents who had left for God knew where. This archaic structure of the land made any meaningful modernization of Juzet's agriculture, dairying, and livestock raising practically impossible.

With these handicaps, the Buzet ethos of heavy borrowing and massive investment in the expectation of spectacular productive improvement never took hold in Juzet d'Izaut. Even the most alert and modern-minded farmers shrank from borrowing, believing, perhaps with reason, that investing in fragile livestock and fancy equipment could do little to raise their net income. They were therefore doubtful about their ability to repay such loans and their existing farm operations did not generate the kinds of savings that would have allowed them to pay cash for the latest equipment. Even such modest improvements as milking machines were adopted late and only by a minority of milk producers: no one was using them in 1970 and only two of five dairy farmers in 1979.[29] Not a single farmer had anything resembling a modern cowbarn or milking shed in 1983.

Declining morale on the part of Juzet farmers, who voiced the opinion that progress had left them behind—*le progrès nous a passé de vitesse*— was reflected and amplified by the reaction of the young, who voted with their feet. In the climate of full employment of the 1950s and 1960s children of farm families moved to Toulouse, Saint Gaudens, and even Paris as factory workers, secretaries, artisans, and postmen. By 1983, an un-skilled farmboy who found a job in the Cellulose d'Aquitaine, the big paper mill of Saint Gaudens, brought home Fr 4,600 a month, 40 percent above the minimum wage, at a time when a married couple of dairy farm-ers, regularly working together late into the night, just might, jointly and after expenses, end up with the minimum wage of Fr 3,200. Parents in Juzet farm families had therefore no stomach for convincing their off-spring to stay on. For instance, when in the 1940s Arrête's boy did well in school, the father pushed him to complete his studies and become a schoolteacher, even though the son would have preferred staying with agriculture. At most, people like Arrête junior and others in civil service jobs with the Forest Service, the Road Commission, or the Post Office might keep cows, steer, or horses on the side.

At least down to the census of 1962, Juzet d'Izaut continued to empty itself of people. From 277 in 1936, population went down to 188 by 1962. Yet the figures understate the exodus, which was partly offset by a return of pensioners to the native village that they had left decades earlier, a pattern we already observed in Buzet-sur-Tarn. In 1962 more than one-third of the residents of Juzet were over the age of sixty. By that year, fifty-six houses stood vacant, including two former farmsteads intermit-tently inhabited by summer people.[30] If population figures have officially risen since then, this owes more to the changing states of mind of Juzet emigrants, than to the village's genuine recovery. Absorbing the perva-sive French mystique of the *résidence secondaire*, the rustic home away from home, these departed Juzet natives have, as we noted earlier, trans-formed most of the abandoned family homesteads into weekend and sum-mer retreats. As a matter of family sentiment, they prefer once more to have Juzet count as their official residence.

The decline in the number of farms was even more precipitous. The sixty-one farm households of 1936 had by 1966 melted to seventeen fam-ilies for whom farming was the principal source of income.[31] Even this low figure was open to question, since by the mid-1960s the inclusion of farmers in the social security system impelled anyone afraid of falling between that system's cracks to enroll as an "agriculturist." When the agricultural census of 1970–1971 provided a breakdown of the nineteen "farmers" it identified, only twelve turned out to dispose of more than 10 hectares of land, five having less than 2 hectares each. Only seven people claimed to be full-time farmers, while of the remaining twelve, eight devoted less than half their time, five less than one-quarter of their time, to farming.[32] Even these figures overstate the case, for the postman's wife tending eight or ten cows would be listed as a full-time farmer, although it was obvious that in this instance the family derived only a fraction of its income from the sale of milk and meat.

Despite its handicaps, in the early 1960s Juzet d'Izaut's prospects as a farming community held some promise. Population seemed to be stabilizing and the village had at least half a dozen young men, mostly in their thirties, who, for varying reasons, intended to remain on the family farm rather than following their brothers, sisters, and cousins to the city. By that time the need to focus on cattle raising or dairying had become obvious. It had become equally evident that keeping seven or eight cows or even ten or twelve was not going to keep a modern farmer going. If they were to expand their herds and their acreage, their obvious priority was to recast the crazy quilt of Juzet's landed property. There was no way of keeping the twenty, thirty, or forty cows needed on pastures intended for two or three. It made little sense to rent or buy additional land, if that land was scattered in quarter-acre lots all over the map.

In December 1964 the municipal council of Juzet d'Izaut, chaired by mayor Hughet, ex-farmer, ex-butter maker turned *charcutier*, voted to request from the authorities the consolidation of the agricultural land of the township. It was not a decision by consensus, as Hughet is supposed to have cast the deciding vote when the council deadlocked. Though Hughet himself still owned his 9 hectares of family land, no longer having the time to farm it himself, he had rented it to a tenant. The mayor could therefore be accused neither of having an axe to grind, nor of being an outsider who knew nothing about farmers and their problems.

The idea had previously been a topic of conversation among some of the young men, including Arrête, who, despite his teaching job, had no intention of giving up the family farm that his parents' hard work and determination had left doubled in size. Yet this was all mere talk, since at the time the whole notion of land consolidation was quite novel in the French southwest and not a single mountain village in the area or, local experts believed, in all of France had yet undergone the experience. The nearest thing was the village of Soueich, southeast of Saint Gaudens, located in the lowest foothills of the Pyrenees, where land consolidation had been begun by the mayor, who also, and not so incidentally, happened to be president of the departmental Chambre d'Agriculture in Toulouse at the time.

Though he had not originated the idea, it was the mayor, M. Hughet, who set about translating it into reality. Recently elected, the mayor was a self-confident man, successful in his business, energetic, and progressive-minded. Like most village mayors he was fiercely devoted to his community. He was also ambitious and what finer monument could one leave than the successful agricultural modernization of one's native village? Hughet was convinced that unless land holdings were made viable by a reapportionment of landed property that would take the present and future needs of modern farmers into account, Juzet was doomed as an agricultural community. None of the young would stay on the farm under prevailing conditions. Land consolidation, he believed, would permit the survival of a few viable, good-sized farms pursuing stock raising or dairying on a scale that had never been dreamed of in Juzet d'Izaut. Their success would encourage the next generation of prospective farmers to stay on in their turn. Juzet would be saved because it would once again have solid economic underpinnings readapted to the modern age.

What was then called the Service du génie rural, that is, the agronomic and civil engineers of the Ministry of Agriculture in Toulouse, greeted the request from Juzet d'Izaut with a mixture of skepticism and interest: with skepticism because, historically, the bureaucratic, legal, and technical mechanism of land consolidation—*remembrement*—had been created and had evolved with grain growers in mind; with interest because *remembrement* among mountain stock raisers and dairymen offered a technical challenge, which, if successfully met, would raise the Toulouse service's standing. By April 1965, mayor Hughet was officially informed that the Génie rural had recommended to the prefect of Haute-Garonne that a land consolidation commission for Juzet d'Izaut be appointed.[33]

Land consolidation was launched, not to be officially completed until February 1972.[34] The procedure was always highly legalistic, bureaucratic, and exceedingly ponderous. What slowed it down even further in the case of Juzet was a combination of political caution on the mayor's part—a good politician will wait for his constituents to catch up with him—and intermittent tight budget years for the Ministry of Agriculture during which the engineers of the Direction départementale de l'Agriculture (DDA) had to cut back on their work.

The task itself was also both difficult and novel. It was unusual to attempt consolidating the agricultural land of a township where working farmers constituted one-nineteenth of the landed proprietors, whereas the remaining landowners were absentees whom the rural exodus had scattered all over France and who had to be notified about what was happening to their land in Juzet. In spite of the letter of the law which assured all proprietors of equal treatment, understandably, land consolidation in Juzet was concerned with improving the farms and working conditions of the minority that had remained on the land. The Juzet emigrants had held on to their land chiefly for sentimental reasons: land consolidation, concerned with creating viable farms, made mincemeat of the absentees' sentiments by reassigning their plots according to the convenience of the handful of remaining farmers.

The technicians were also up against a pattern of miniplots that has already been delineated: usually a township field plan is drawn up to the scale of 1 to 5,000. In Juzet it had to be blown up to 1 to 2,500, twice as detailed as a standard field map, so that the smaller fields would show up at all. Even then some were too tiny to be labeled. The engineers and surveyors tried to create sizable fields, ending up with an average size of 3 hectares. They also sought to integrate the land that a farmer rented with the land that he owned. In some instances, the disillusioned absentee proprietors offered their land for sale, rather than seeing treasured family fields wander all over the landscape. This naturally helped the process of farm enlargement that consolidation was intended to encourage. Enclaves disappeared. Abandoned fields of uncertain ownership were moved to marginal land on the periphery. Municipal roads were asphalted, field access roads meant for modern equipment laid down. And since it was generally agreed that Juzet would depend on cattle, every field was provided with a drinking trough fed by a brook-fed watering system.

Although, as usual, farmers might disagree as to who had gained most and who had been shortchanged, most of the younger farmers could agree that land consolidation had revolutionized what in retrospect some called the "inhuman" conditions in which they had previously labored. *Remembrement* in Juzet opened up the possibility of adapting local farming to the modern age. The engineers from the DDA had optimistically predicted that from the then current 370 head of cattle, the local farm economy would grow to 600 head in the foreseeable future that land consolidation had ushered in.[35] Individual dairy and meat cattle herds did become larger in succeeding years, but overall numbers declined, as old farmers retired or died without full-time successors. New part-time farmers were satisfied with a less intensive use of the land demanding less of their time. The agricultural survey of 1979 counted only 146 head of cattle in Juzet d'Izaut, numbers that may have risen slightly since then.[36] And this decline took place, despite the fact that two of Juzet's farmers were once again using the Cagire for summer pasture, a practice abandoned in the postwar era. Tractors and devices such as fertilizer pressure sprayers did become generalized and several farmers even created an equipment coop to share such fancy equipment as a mechanical manure spreader. Wheat and corn fields planted only because a given plot was too small to be used as meadow or pasture became a thing of the past.

And yet despite its technical success, land consolidation has not saved Juzet, but merely bought time: twenty to thirty years of continued farming, according to Mayor Hughet himself. The results were good enough to keep going the generation of farmers who had pinned their hopes on *remembrement*, but not good enough to convince their sons that mountain farming had a future. Once the youngsters had made their intentions clear by becoming electricians, road equipment operators, postmen, and clerks, the morale of the community was broken. Ironically, even if land consolidation had failed to keep full-time farming viable in Juzet d'Izaut, it at least eased the agricultural community's transition to the next phase of its decline.[37]

Just what has happened in Juzet d'Izaut over the last decades may best be understood by looking at the careers of three farmers, all of them near or in their fifties at the time of my interviews, who personify the range of local problems.

All his neighbors agree that M. Faulin is a very hard worker, *"il bosse, c'est un bûcheur."* The lights in his cowbarn are often still on at midnight. He is also respected in the community as the model full-time dairy farmer who has built up his farm and his herd, kept abreast of progress, and who, at least as long as he does not start counting the hours he and his wife put in, makes a decent income. Faulin himself did not share this cheery view.

Faulin's father and grandfather had been typical Juzet peasants except that they had also run the now long-defunct milk plant, which supplemented their farm operation, providing them with a good living. People expected less then, Faulin emphasized, and once they had paid their taxes, they were free and clear. They were not weighed down with *charges*, with mandatory social security payments, however beneficial social welfare provisions may be. Faulin himself, who was about to turn fifty at the time he spoke to me in 1983, had gone to work on his father's farm of 8 hectares at the age of fourteen, but had not taken over on his own until he married when he turned thirty. This was an era when the most prosperous peasants, owning 10 or 12 hectares, had ten to fifteen cows, while most made do with five or six. At that time, he and his young wife had agreed to nurse an elderly maiden lady left alone in the world, who, upon her death in 1965, willed them her family holding, 20 hectares, which had previously been worked by sharecroppers. Such arrangements, which we will encounter in other villages, variants of what the

French call purchase by *rente viagère*, are fairly common in the country-side of the French southwest, though the provisions of the French inher-itance laws tax unrelated heirs very heavily. Faulin was also able to round out his property by buying between 2 and 3 hectares of additional land and by leasing another 5 or 6. With 36 hectares of land at his disposal, Faulin's farm is among the three or four largest operations in Juzet d'Izaut.

When Faulin took over from his father, he had inherited fifteen head of cattle. He has built up the herd to thirty-six, calves, heifers, and bul-locks included, of which fifteen are milk cows, the rest raised for meat. Though milk is poorly paid and ties you down far more than raising steer, for someone like himself who lacked working capital and whose turnover of animals sold for meat was bound to be slow, milk was indispensable: at least this way he could count on collecting some Fr 5,000 every month, though the net is much smaller considering the high production costs of milk. He sells his calves and steer to wholesale butchers, though he re-gretted the virtual disappearance of the seasonal cattle markets at Saint Gaudens and elsewhere. Cattlemen coops sound like a nice alternative in theory, but in everyday life they suffered from excessive overhead and were slow in paying for the animals they marketed. Because he got by, but with difficulty, he avoided the Crédit Agricole as much as possible. Five years ago, when he was entitled to low interest after a drought, he did take out a loan to redo his roof, which he expected to finish repaying in 1983.

Faulin has pushed the modernization of his farm as far as he has dared. For several years he worked with the agricultural extension agent, but felt that the kind of advice he was receiving was intended for large-scale farmers, who could afford heavy investments. He did become convinced, however, that grass silage, if fed to milk cows in moderate quantities, was a good thing. Ever since 1978, with the help of a cantonal Coopérative d'utilisation de matériel agricole (CUMA) that has acquired the necessary equipment and which organizes mutual help among its members, Faulin put 3 hectares of grass into silage, the only one in the *commune* to do so. His land carried one cow per hectare, which was a more intensive land-use than that of his neighbors. He was well aware that he could increase his herd by doubling the five or six tons of chemical fertilizer he sprayed on his meadows, but, though he did not spell it out, it was obvious that he saw little point in so doing when he was within ten or fifteen years of retiring without a successor to take his place. In any event, he was leery of laying out more money and he was already working long hours. He was also one of the last two farmers in Juzet d'Izaut to regularly plant crops.[38] In spite of the inroads of deer and boar, he grew corn, potatoes for the family, and a field of barley, which he milled himself and that, in the form of flour, supplemented the soy cakes he had to buy for his milk cows. Because so little land was planted to crops in Juzet, it was difficult to get a harvesting contractor to come up there with a combine. If it weren't for the fact that he was on friendly terms with a contractor who brought his machine just to help him out, it would have been a real problem. As it was, Faulin has been forced to harvest his corn by hand.

Yet in contrast to the high-powered dairymen of Buzet-sur-Tarn, Faulin had not invested in up-to-date barns and milking sheds. He felt lucky to be able to afford second-hand, portable milking machines in the early 1970s and, more recently, the refrigerated milk tank that the dairy company picking up his milk has come to require. Nor did he feel he could afford to buy semen from prize bulls or super-cow embryos for implantation: such things took place in a world that, as far as Juzet d'Izaut was concerned, might just as well be in outer space. Faulin's milk cows continued to be bred by a local bull. At an average of 2,700 liters per year, their milk production was almost double that formerly furnished by the gray Saint Girons breed that has vanished from the village, but the best Buzet-sur-Tarn dairymen squeeze out two and one-half times as much from their overbred beasts.

By dint of persistence and hard work both on his part and on that of his wife, Faulin gradually accumulated the equipment he needed, beginning with a tractor in 1969, the second one in the village, which he acquired just before land consolidation was completed. *Remembrement*, for which he gave full credit to the mayor, also helped out a lot: the 112 dispersed fields that he owned before have been reduced to four or five large chunks of land that are much easier to work and more accessible. As far as his equipment was concerned, everything had been bought second-hand, and some of it, he realized, was obsolete, like the baler that neatly compresses hay into cubes, where the modern machines seen in the plain roll the hay into huge water-repellent rolls that evoke gigantic snail shells. He and Arrête have organized the local CUMA already mentioned, through which three Juzet farmers now share equipment used only from time to time, like manure spreaders, that they could not afford individually. On the other hand, he was in something of a panic about having to replace his aging tractor, having no way of paying for a new one: at 10 or 12 percent interest, he did not see how he could afford to borrow.

Where most farmers in Juzet were embittered, Faulin was ambivalent. He could have joined the Forest Service as his brother did, but he enjoyed being a farmer and liked the life. Yet he could understand why his only son would have no part of it. The government should have done something, Faulin believed, to provide incentives for the young to stay. He did not blame his son, who never did like farming and who hoped to return to his job as electrician as soon as he was through with his military service. The father had to admit that, as things stood, to rely on farming for a living in Juzet d'Izaut was impractical. The future, if any, belonged to those who had some sort of job on the side. When not long before, the government had raised the French minimum wage to Fr 20 an hour, he realized how far he and his wife were from earning even that minimum for *their* labor. And when he thought of the tractor he could not afford to replace, it angered him to realize that after all these years of hard work— fifteen and sixteen hours of it during the summer—he had nothing to show for it. Even though he still liked what he did and intended to keep working until retirement, there were times when he could not help but be discouraged.

Where Faulin was aggrieved, his friend Dancausse was eaten up by his grievances. Dancausse really felt beleaguered, in spite of the fact that he had something that no other farmer in Juzet could claim: a son who, within two years, would be ready and able to take over the family farm. Even though Dancausse referred with great formality to M. Hughet as *mon cousin, Monsieur le maire* and was more or less related to all the other families established in the village, he and his wife were nonetheless outsiders: not only physically, as the only family whose farmstead is outside the village boundaries, but in other ways as well. Dancausse himself was the grandson of a Canadian who by some quirk of fate had married a local girl and settled in this remote Pyrenean village; Dancausse's wife was a more recent newcomer: educated and city-bred, she was the daughter of repatriated French North Africans, who only became a farm woman by marriage. The Dancausses, moreover, were the sole sheep farmers among dairymen, and cattle and horse raisers. Nor were they people to suffer in silence.

Dancausse's father's farm of 25 hectares had been the largest and most prosperous in Juzet d'Izaut: at its peak in the immediate post–World War II years, the farm had been stocked with 120 ewes, 22 head of cattle, and 1 mare from the time when a horse and buggy was the only alternative to an ox-cart or to walking if one wanted to get to Saint Gaudens. Much of the Dancausse land had been in crops of all sorts, grown and harvested with the regular help of local day laborers. When in 1961 the mother had died, the father, vigorous despite his sixty-six years, refused to turn over the farm, which still had eighteen cows and thirty ewes, to his thirty-seven-year-old son. In the ensuing family squabble, the son and daughter-in-law left the homestead. The young Dancausse went to work as a lumberjack on the other side of the pass, not to return until years later, when his father sent for him to take over. By that time, the enfeebled father had run the family farm into the ground: all that was left in the way of livestock were a pair of cows and thirty sheep. In retrospect, Dancausse was not sure whether he did the right thing in returning to take over such a nearly bankrupt farm. For a while, the Dancausses concentrated on raising calves, but in 1972 decided to switch to ewes, much of their land consisting of steep hillsides best used as sheep pasture.

By 1983 the Dancausses had 130 ewes. Originally they had crossed the local race, the *Tarasconaise*, with a Suffolk ram, which produced a hardy hybrid that could thrive outdoors summer and winter, whose sturdy and well-fleshed lambs were much in demand among the wholesale butchers to whom they sold. Recently they had been forced to change by breeding their ewes with the *Laconne* variety from the center of France, which produced sheep that were lighter and more agile, less likely to hurt themselves when panicked, but with considerably less meat. The Dancausses' farm income consisted of the proceeds from the sale of male lambs—which fetched an average of Fr 250—and by selling off ewes that no longer lambed. Wool did not count: it sold for Fr 2.50 a kilo, when contractors asked Fr 4 for shearing a sheep that had only 700 grams of wool. Yet when his daughter who lived in Paris had a mattress stuffed, she paid Fr 75 per kilo of wool. They supplemented the income from the sheep by fattening a pair of oxen. By the most optimistic estimate, their figures pointed to a *gross* yearly farm income of Fr 45,000 or Fr 3,750 per month.

Even though they had bought land, enlarging their farm to 45 hectares, the largest in Juzet d'Izaut, 130 were not enough to make a living. It would take, they assured me, three hundred ewes and for such a flock they lacked adequate pasture, both in quantity and quality. Much of the land they had purchased was in woods and if they managed to make ends meet at all—and their country kitchen bore the unmistakable stamp of rural poverty—it was by continued lumbering on their own land. Dancausse systematically felled mature chestnut trees for sale to a small Aspet firm that made hardwood flooring.

What aggrieved the Dancausses, first of all, was their feeling of being unable to compete with part-time farmers who, as employees of the Road Commission (there were no less than four of them in Juzet), as postmen, as teachers, could rely on their monthly salary: *Ils rentrent leur mois* was their resentful refrain. By contrast, they themselves had perhaps been foolish enough to feel that farmers should not take jobs away from city people who had nothing else. Like the Faulins and a couple of others, they were therefore dependent on what they could make from farming. Though most of their land had been located outside the perimeter of land consolidation, *remembrement* had hurt them indirectly, because the owners of land that they had previously rented by verbal agreement had used land consolidation as an excuse to renege and take back their land. Yet when the postmistress of Izaut l'Hotel had inherited her parents' farm in Juzet a few years back, she had not only taken over the land they had owned, but the land they had leased as well. She had land enough for grazing horses and horses take a lot of land. It was to be expected that land would be unavailable for rental in all of Juzet when part-timers were allowed to compete unfairly with full-time farmers like themselves who had a living to make. What added insult to injury was that these people could afford powerful tractors and fancy cattle-feeding installations, when real farmers did not know which way to turn. The others could always count on their monthly salary check. And in the case of the school-teacher, his wife's as well.

They were even more embittered by their conviction that in Juzet d'Izaut farmers were being sacrificed to hunters. The depredations of deer and boar had forced the Dancausses to abandon growing crops. They had tried everything to frighten away the game, even firecrackers, until the people in the village had complained about being awakened at night. They were not alone: one of their neighbors had become known as "Monsieur France-Inter" after the all-music radio station, because he had tried frightening game with a transistor radio. Nothing worked after the first few days: red does ate the corncobs, roe deer trampled the rye, boar uprooted potatoes and beets.

There had always been game in the area, but the large red deer were introduced about the time of land consolidation, since the powers-that-be had evidently decided that Juzet was good only for meadow and pasture. And most of the wild pigs had been raised in captivity and released to keep the hunters amused. Last year the hunting association had bagged fifty-seven of them and some years they had reached one hundred. When one of their neighbors had finally succeeded in getting the game inspector to come up to Juzet, the inspector could not believe the number of deer he saw in the handful of remaining fields. If there were no more deliberate efforts to increase the density of the area's game and if all remaining farmers agreed to plant some of their land to crops to spread the risk of depredation by deer and boar, real agriculture might once again become feasible. As it was, in 1982 they had to buy 2 tons of corn in the plain for fattening their oxen, when, if it had not been for the game, they could have grown it themselves. What was worse, they had to sell scrawny little lambs, because, unlike sheep growers in Izaut l'Hôtel who were not so close to the forest, the Dancausses could not raise their own barley to fatten them. They would have liked to go back to raising the yearly family hog as well, but having to buy feed on the outside made this impractical.

That was not all. The hunters and their dogs were driving the Dancausses out of business as sheep farmers. In eleven years, they had lost seventy sheep to them. Posting their land did no good whatever. Once the pack was released, the dogs ran wherever boar or deer led them and they did not make any fine distinction between game and sheep either. He had found sheep killed by dogs, animals half eaten, ewes that had been chased until they broke a leg. That was why they had been forced to switch to a lighter breed of sheep, even though the new variety was much less profitable. The Suffolk were too heavy and more likely to hurt themselves in trying to get away from the hunting dogs. He had tried to get Arrête, the schoolteacher, organizer of the local hunting association, and keeper of its pack of dogs, to give him a couple of days' notice before a hunt, so that Dancausse could gather in his flock. Arrête and the hunters contended that they had no way of knowing in which direction the pack would take off. Even the Dancausses' own interest was unclear. During the fall hunting season, they relied on their chestnut tree grove for fattening their sheep: in a couple of weeks, the animals gained Fr 10,000 or 15,000 worth of weight. Yet if the flock were removed from the grove for a day or two, they would eat themselves sick with accumulated chestnuts once they returned. The sheep had to keep pace with the falling nuts to avoid trouble, so moving them about at all at that time of year was risky at best.

When Dancausse has complained to his cousin, the mayor, who was also, after all, vice-president of the hunting association, he had not received much sympathy. When he went to show the mayor a sheep that had been literally eaten alive, with a good pound and a half bitten out of its upper leg, the mayor told him, "You make me shit, you and your stories of hunters and sheep!" Dancausse had even found one that had been shot, though sheep were hard to mistake for either deer or boar and he was sure that the many sheep that had vanished altogether were shot and stolen by hunters. He has had confrontations with hunters who were going to follow their dogs into the last unmowed square of a meadow which he, Dancausse, was in the process of mowing. In 1981, in separate incidents, both he and his wife had been shot in the back, he by a bullet, she with a load of buckshot. The public prosecutor had never found a suspect. His cousin, the mayor, Dancausse learned recently, had informed the authorities that the sheep farmer was not known to have enemies.

In the meantime the Dancausses were grimly hanging on until their son could take over in two years' time, once he graduated from agricultural high school in Saint Gaudens. What he would do then would depend on the agricultural councillors' advice and on how he used the government aid that young farmers receive when they start up. As another Juzet farmer had pointed out sardonically, it was just enough to pay for about two tractor wheels. Mme Dancausse was convinced that her son would have to have some other source of income on the side to be able to make it on their farm. They didn't wish it on him to have to struggle the way they did, just to keep his head above water. In anticipation of his son's taking over, Dancausse, who had been all over asking for advice, has decided to enclose his sheep grazing land, all 3.5 kilometers of it, even though it would cost Fr 70,000. He had already bought the first kilometer of fencing. A fence should keep out the dogs and, if a boxwood hedge was planted just inside it, the game as well. He was also hoping to get a caterpillar type tractor with which to plow some of the steep sheep pastures, reseed them, and systematically renew the grazing, so that their land would support more sheep. There was even a sizable level area on the crest of their hill, where, he was convinced, 4 or 5 hectares of crops could easily be planted once the incursions of game were stopped. Dancausse had not quite given up on farming, still hoping that his son would be able to make it. When other farmers in the village were asked about young Dancausse's prospects, they shrugged their shoulders. But then again, since the Dancausses, who were not well thought of in the village, were the only farmers in Juzet d'Izaut to have a son to succeed them, local reaction may have been tainted by envy.

Whether or not young Dancausse succeeded, Arrête, the school-teacher, was already a success. He was, if there is such a thing in Juzet, the wave of the future, the man who collected a salary, yet who had managed to do something with his family farm. Arrête had the happy mien of a man who expected applause from his fellows and usually got it. He has been as successful as anyone in the community in combining the rewards of the great outside world, embodied in this instance by the Ministry of Education that pays his monthly salary, and the local standing that naturally comes to those shrewd enough to make something of their family farm.

His father had already been a progressive hard-working dairy farmer, who in his lifetime had doubled the size of his farm to 15 hectares and switched to the more productive breed of Swiss milk cows. The father was known in the village as someone who was all work and no play, yet not wanting his son to have the same hard life he himself had led, pushed him to continue his education and go into teaching. The son would have preferred staying in farming and was convinced that had he remained a full-time farmer in Juzet, he would have proved resourceful enough to have found some way of complementing the livelihood gained from agriculture. It took some gimmick on the side, *un truc à côté*, he maintained, to make a decent living in a place like Juzet d'Izaut, maybe a cheese factory. He had a friend, for example, who built himself some chicken coops and when the prices of ordinary eggs went down, he specialized in fertile eggs for hatching chicks. The important thing was to remain flexible and alert to opportunities.

Arrête junior had continued the family tradition of enlarging their farm by investing the savings from his and his wife's schoolteachers' salaries in land. Like his father before him, he too has doubled the acreage of the family farm; he now owns 35 hectares and rents another 5 on the side. Twenty years earlier, he had been one of the first to talk up land consolidation for Juzet, though at first such an idea was not well received, not even by his own father, who was on the municipal council at the time. When the mayor became an advocate and *remembrement* was launched, Arrête junior saw his opportunity. By dint of lobbying, persistence, and a willingness to accept even poor land in exchange, he persuaded the surveyors and the subcommission in charge to let him have almost half of his land as a single lot: 15 hectares that he could fence in and use as pasture.

In the ten years since land consolidation had become a fact, Arrête had managed to create a cattle-raising operation that he could run with a minimum of work, now that his father had turned eighty and the old man's energy declined. On his 15-hectare lot the son had built an open stable with an ingenious self-service manger of which he was very proud. He merely needed to fill it up with hay on Wednesdays—the weekly French school holiday—and Saturdays. His cattle roamed and grazed freely in their vast enclosed pasture, returning to their stable and manger when they felt so inclined. In order to improve the quality and quantity of their meat, Arrête had crossbred his father's Swiss cows with the large, square Limousin breed that adapt well to the rugged climate. Even though the animals were outdoors all year, they were in excellent health. So far he had been keeping and raising all of the female calves, only selling overage cows and one-year-old bullocks. Yet with nearly forty head of cattle, his herd had reached its upper limit. His pasture land was treated with manure; his natural meadows—aside from the inconvenience of extra work, in Juzet artificial meadows like clover, alfalfa, or *luzerne* were no longer perceived as paying propositions—were regularly sprayed with chemical fertilizers. He had bought the equipment he needed, including a 70 horsepower tractor, the most powerful in the village, and the *commune*'s only baler that made the huge hay rolls that he needed for his self-feeding manger. The manure spreader and the hydraulic fork he shared with Faulin. He was all set.

Arrête made no bones about the fact that for ten years he had been pouring money into this cattle-raising operation without much return. He also admitted that if, suddenly, he had to rely on nothing but the farm income that his capital and effort had built up, he would end up with less than the legal minimum wage. Yet for him that was beside the point: since schoolteachers in France may take their retirement at fifty-five years of age, he himself was within two or three years of retirement, which was why he was building himself a villa just up the street from the family homestead where his parents still lived. Once he retired to Juzet, he would be where his roots were. For the first time in his married life, he would be moving from Spartan schoolmaster's lodgings in a rural schoolhouse to his own home right next to the family farm he had built up and where he also kept his hunting association's dog pack, an agreeable year-round reminder of the excitement of the fall hunt. At the same time, with the two or two and half days of work a week he had put in all along, his herd of cattle would nicely supplement his and his wife's pensions: they would make out as well in retirement as they had during their professionally active lives. And if, once they reached adulthood, one of his three children were to show a taste for agriculture, he, their father, was making sure that there would be an inheritance worth taking up. Who could predict the future in this day and age anyway? He prided himself on being someone who kept his options open.

3

Changing Course at Saint Cézert

"SAINT Cézert," I was told by a technician who knew the area, "is on the road to nowhere. Nothing much goes on there. *Ils sont tranquilles.*"[1] The comment had a familiar resonance. Almost one hundred years earlier, Jean-Marie Sapène, the village schoolmaster, had noted in his official report: "The country road going from Burgaud to Grenade is the only one in the *commune* that will accommodate a carriage. There are no railways, no stage coaches, no fairs, no markets: because of its geographical location and its insignificant population, Saint Cézert is doomed to isolation."[2]

Some one hundred years later, the schoolmaster's poignant description needs only a slight updating. The township's population of 180 people is half of what it was when he drew up his report and the village itself is down to less than 100. Although Saint Cézert is only 21 miles northwest of Toulouse, there is no trace of encroaching suburbia, you find none of the villas of pastel-colored concrete that are scattered throughout the Buzet-sur-Tarn countryside, for example. Nearby Grenade, 5 miles closer to Toulouse and a busy market town on a well-traveled departmental road, constitutes suburbia's cutting edge. At least so far, Toulousains have found Saint Cézert too distant and too isolated for suburban building lots. A number of Saint Cézert residents of long standing do commute, including the mayor, who is not in good health and has traded active farming for a desk job at the Chambre de'Agriculture. Yet the little village, with its hole-in-the-corner grocery, the garage where old men stand around watching a tractor being repaired, the few children playing desultorily by the church, the empty square baking in the midday sun, still conveys the same impression of being way off the beaten track. The sole carriageable road the village school teacher mentioned back in 1886 has long been asphalted, but, more often than not, a half hour will elapse between passing cars. Saint Cézert is indeed on the way to nowhere.

Geographically, Saint Cézert is an ancient tableland, trisected into three "islands" of very unequal size by valleys carved out by the two brooks that cross the township. These two streams—the Margesteau being the more important of the two—enter the *commune* from the southwest and northwest respectively, join in its center, and leave as a single, small creek by the northeast. The U-shaped valleys the streams have dug

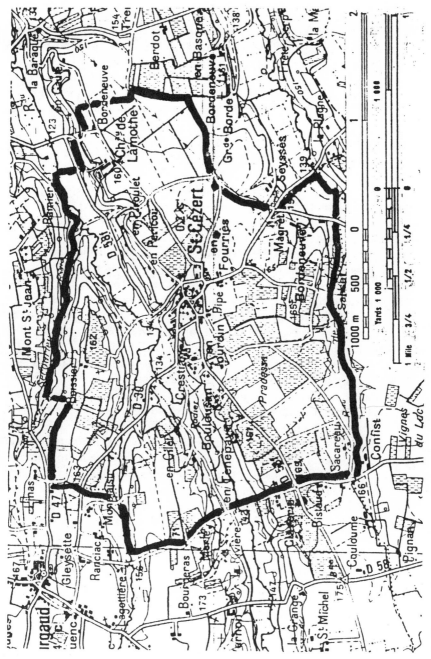

9. The township of Saint Cézert

link Saint Cézert to the wide plain of the Garonne River to the east, while the modest highlands that dominate the rest of the township are part of the first terrace of the Gascon hills to the west. The river-bottom land, less than 20 percent of the *commune*'s acreage, is grandly known as *la rivière*, the river. To the south of *la rivière* is the low plateau, comprising some three-fifths of the township's acreage, where the *bourg* of Saint Cézert itself is implanted. This area is locally referred to as the plain (*la plaine*), supplemented in the northeastern quadrant of Saint Cézert by two much smaller extensions of highland terrain similar in elevation, topography, and soil characteristics. The transition between "river" and "plain" consists of a strip of slope (*les côteaux*) varying in steepness, a grade that at various times has comprised woods, vineyards, and sheep pasture. Nowadays woods have largely taken over, some 100 hectares of them.

Before recent irrigation schemes undercut long-established judgments, *la rivière* was considered Saint Cézert's most valuable land. Much of it is heavy alluvial soil that remains moist throughout the summer and could be cropped year after year without fallowing. This was also the area where the *commune*'s natural meadows were concentrated, which is why any farmer with draft animals had to have access to some land in *la rivière*. Moreover, it was the only soil in the township where corn could be grown with some assurance of success.

La plaine and the analogous highland sections north of the Margesteau brook consist of unpromising soils that recall some of the worst land at Buzet-sur-Tarn. Here and there patches of fertile, water-retaining clay can be found, but most of the plateau is divided between gravel and *boulbène battante*. The former is infertile but well-drained. Given the sunny climate, it has potential for growing grapes. *Boulbène battante*, on the other hand, combines low fertility and an extreme sensitivity to summer drought with a high vulnerability to winter rains, which the cement-like substratum tends to trap until well into the spring, very much the same sort of soil conditions that bedeviled most of Buzet's farmers. In Saint Cézert the problem is compounded, because during spring and summer, the upland is also exposed to the region's southeast wind (*le vent d'autan*) that in a week's time can dessicate any crop. Given the extreme contrasts in the character of the soil, most Saint Cézert farmers of substance have traditionally owned land all over the township, unlike the more limited dispersal at Buzet, for example.

Compared to Buzet or to Juzet d'Izaut, what is striking about the character of Saint Cézert is the much greater historical continuity of its social structure. For instance, the domain of Saint Cézert's prerevolutionary lord, the count of La Mothe Descars, a domain taking in almost one-fifth of the *commune*'s acreage, remained virtually unaltered as the major local estate for two hundred years, until it foundered and was broken up a dozen years ago. Although it is true that the estate of La Mothe changed hands during the French Revolution, its new bourgeois owners carried on the eighteenth-century tradition of absentee ownership and management by way of bailiffs supervising *maîtres-valets*, plowmen lodged by their employer and hired by the year. Despite its vast size—and it spilled over into adjoining Aucamville—La Mothe never came to dominate Saint Cézert. At best, most of its absentee owners took only an intermittent interest in the *commune* in which their property was partly located.

The people who have dominated Saint Cézert over time, who have owned the largest share of the land, who have furnished most of the mayors, are a group of substantial farmers. This is not to argue that ever since the eighteenth century, the same biological families have remained within this charmed circle. Local marriage alliances complicate the random pattern of succession and extinction, of good fortune and hard luck, of shrewdness and incompetence, that make village genealogies so hard to disentangle. In any event, there is no overlap between the names of the eighteenth century's major landowners and the handful of today's substantial farmers whose roots are local. Over the last 150 years, to judge by the land register drawn up in 1830, the turnover of names has been less complete: several of today's prominent families were already listed as substantial proprietors in the 1830 cadastre. Yet whatever the biological continuities, some such group has been in existence at least since the eighteenth century and perhaps earlier. If one consults the eighteenth-century tax assessment rolls (quaintly titled *martyrologues* or list of martyrs) one finds just below the assessment for two great aristocratic estates in Saint Cézert (which together paid 23 percent of all local taxes), six landowners assessed at 25 to 50 *livres* annually and a further ten in the 10 to 24 *livres* range. In contrast to Buzet-sur-Tarn, where most such landowners were clearly identified as either nobles or bourgeois, in Saint Cézert the taxpayers assessed between 10 and 50 *livres*, who as a group paid 40 percent of the taxes, seem to have been more like sixteenth- or seventeenth-century England's solid yeoman farmers.[3]

Just as yeoman farmers were never quite gentlemen, these Saint Cézert farmers were not really bourgeois, though there have been times when the lines became blurred. The 1830 census, compiled when one of their number served as mayor, does not even distinguish them from their own hired hands: both were indiscriminately labeled *laboureurs* (plowmen) as compared with *brassiers* (day laborers), who usually doubled as smallholders.[4] Like their English counterparts, these French farmers also hired agricultural laborers to do what heavy work needed to be done, but when it came to official documents like census descriptions, they evidently preferred to err on the side of modesty.

This official public shyness should fool no one. Certainly from the period of 1830s on, the evidence points to the existence of an agrarian elite in Saint Cézert that not only owned a disproportionate share of the land, but directly controlled much of the local population by virtue of its economic power. From 1830 to 1948 (a date that coincides roughly with the end of premechanized agriculture in the township), nine or ten proprietors owned all farms above 20 hectares, that is to say, by traditional standards farms that were definitely too large to be worked merely with family help. Jointly they never owned less than 54 percent of the total taxable land. In 1948, their share was as high as 63.2 percent.[5] When Jean-Marie Sapène, the schoolmaster (and in his own right a shrewd investor in local farm real estate) sought to describe the community, he had no doubts as to how things were run. After mentioning the fully consolidated domain of La Mothe, owned at that time by a Monsieur Monge who had made it into a model farm, Sapène wrote: "The rest of the *commune* is therefore parcelled out among very well-to-do landowners, each one of them at the head of his workers. For the most part, these proprietors play the leading role in the cultivation of their land."[6]

Two decades, later, in 1906, an unusually informative census of Saint Cézert permits us to sketch more specifically the patriarchal society to which the schoolteacher had alluded: the four top families of proprietors, the Davasse, the Mas, the Touzoulé, and the Guichot, employed 42 male workers, who, together with their dependents, comprised 136 people out of 300 local inhabitants, or, more to the point, out of an *agricultural* population numbering 259. Five other landowners employed an additional 11 adult males, most of them with families as well.[7] Altogether some 170 of the townships's residents (and two-thirds of its farming population) were dependent upon nine substantial employers. Even these statistics may understate the oligarchy's influence: they ignore the handful of smallholders who needed intermittent casual employment to be able to survive. This overall pattern of a community sharply divided by class into employers and employees, landowners and landless or land-poor, was to outlast World War II. Since the 1960s, as we shall see, the pattern has partly eroded, to the point where in 1983 there were only four salaried farm workers left.

This dominance of well-to-do yeoman farmers employing the bulk of the population has had considerable influence both on Saint Cézert's demographic evolution and on its agricultural development. In Juzet d'Izaut, throughout the eighteenth and nineteenth centuries, we observed the struggle for survival on the part of a peasantry whose numbers had outstripped resources. In Buzet-sur-Tarn, even though more than half of the land was owned by urban absentees, the struggle may have been less desperate, because the population pressure was less intense. Nonetheless, until late in the nineteenth century, subsistence—growing the food required by each peasant family—remained the crucial issue for most sharecroppers and small owner-operators. In Buzet, some peasants were only partly relieved of this dependence by the boom in green peas in the wake of the coming of the railways and the growing market for local table wine.

The fact that Saint Cézert was a rural community dominated by a farming oligarchy influenced its demographic evolution, though the exact shape of that influence is not so obvious. What is evident is that the demographic history of Saint Cézert from the end of the eighteenth century to the 1950s differs notably from that of communities like Buzet-sur-Tarn or Juzet d'Izaut. Admittedly, the general regional population pattern—an increase to the middle of the nineteenth century, followed by a long gradual decline, sharply accentuated by World War I—also has some relevance for Saint Cézert. What makes Saint Cézert different, however, is that these general trends were at various times comparatively muted or accelerated. The population increase from the 1790s to a high between 1831 and 1861 was a moderate 25 percent, from around 340 to between 415 and 440 inhabitants. Yet, between 1861 and 1871 population dropped sharply by slightly more than a quarter, to hover between 300 and 340 for the next forty years that ended with the outbreak of World War I. By 1921, the population had fallen another 20 percent to 244, but unlike most other purely agricultural communities, the number of inhabitants stabilized and then crept up. With a population of 271 in 1954, numbers were up 10 percent over what they had been in 1921. In the absence of suburbanites, owners of secondary residences, and home-sick pensioners, the *commune's* population has been dropping rapidly since then—down to 180 by 1977—but that is another, more universal, story in which Saint Cézert has been swept up by national and even international trends.[8]

What accounts for Saint Cézert's distinctive pattern of population growth and decline? In the case of Juzet d'Izaut, we saw crude Malthusian checks at work to curb growth beyond a certain point: after 1850, a reestablished balance between local births and deaths went hand in hand with emigration, some of it to the New World. Prior to World War II, the stability or decline of Buzet's population depended on the relative attractiveness of wine and vegetable growing on one hand, city living and urban jobs on the other. It is true that in both cases decline may have been accentuated by the Southwest's general trend toward smaller families. Although Saint Cézert was by no means exempt from this trend, there the whole mechanism of population dynamics differed in being controlled by the *commune*'s employing farmers.

Once again, comparison is called for. However contrasting the character of farming and of peasant life may have been in the two communities, in both Buzet-sur-Tarn and in Juzet d'Izaut, inheritance and succession made the difference between stability and decline. A son's refusal to take over his father's small farm (among big farmers the issue rarely arose until very recently) was really a decision to emigrate, as was a daughter's refusal to follow in her mother's footsteps by marrying a local farmer. By staying or leaving, young people unwittingly took part in a referendum on their home community's survival. Whatever the pressure of circumstances, such decisions to assume or not to assume traditional roles were individual and, ultimately subjective decisions.

Until very recent times, this was not how things were decided in Saint Cézert. The great majority of the local population down to the 1950s were neither peasant proprietors nor tenant farmers, but agricultural workers hired by the year, supplemented during the nineteenth century by casual laborers (*brassiers*) who usually owned a couple of hectares of land that they worked by spade. Therefore the crucial question that determined the growth, stability, or decline of the local population was not so much the question of succession, but the number of available jobs. If, as seems to have happened between 1861 and 1872, the yeoman farmers introduced labor-saving devices, laborers thrown out of work had no choice but to leave with their families.[9] We may speculate that it was harvesting equipment that was introduced in this particular instance, reapers or threshing machines or both, and that the victims of technology were Saint Cézert's *brassiers*, rather than its permanent farm workers. *Brassiers* had constituted 29 percent of the local population in 1805, about 25 percent in 1830, yet they were down dramatically to 8.5 percent by 1906. As Sapène first shrewdly pointed out, in Saint Cézert population decline was an indication of technological *progress* rather than of social decay.[10]

The extent to which employing farmers were able to counteract "natural" population trends shows up in the changing composition of the labor force during the interwar period and even beyond. As elsewhere in Haute-Garonne and in France generally, World War I not only killed a number of men in their most productive years—there are nine names on the local monument to the dead of World War I—but it speeded up the exodus of the rural population to the cities. Yet local plows continued to need plowmen and vines still had to be trimmed: the need for hired labor was not going to abate just because some of the demobilized soldiers chose Toulouse over Saint Cézert. Even though the rural exodus continued, between 1921 and 1936, population rebounded from 244 to 262.

The answer to this puzzle was that outsiders replaced departing natives. There had not been a single foreigner in Saint Cézert in 1906. By 1921, the census listed ten Spaniards, presumably men who had been brought in to work in the French countryside during the rural labor shortage of the war years, a shortage that the end of the war failed to make good. By 1936, Saint Cézert's employers had attracted—temporarily, as it turned out—a sizable colony of foreign farmworkers and their families: the foreign-born, almost all of them Italian by that time, had grown to sixty-one, nearly a quarter of the township's inhabitants.[11] After World War II, the Italians either found better-paid work in the city or left to become tenant farmers elsewhere. In their turn, they were replaced by Muslim immigrants from Algeria and Morocco, for the farmers of Saint Cézert still employed thirty-nine hired hands as late as 1954.[12] Local demand for farm labor continued to shape the demographic profile of the community: it was as if the big farmers could simply go out into the marketplace and hire whatever replacements it took to keep their vines trimmed and, more incidentally, their village populated.

The men who owned the bulk of the land in Saint Cézert also left their mark on the agricultural history of their village. If we define a peasant as an agriculturist who puts self-sufficiency ahead of profits, the yeomen of Saint Cézert were never peasants. And that social fact did make an agricultural difference. Though the prompt adaptation of Buzet agriculture to the railway age should warn us against facile clichés about peasant traditionalism, the landowners of Saint Cézert, accustomed to producing for the market, were likely to be even more responsive to its indicators. It is true that well into the twentieth century, Saint Cézert residents depended on food grown in the township, yet in the light of a ratio of between 1.5 and 3 hectares of arable land for every inhabitant (in Juzet d'Izaut it averaged as low as .5 hectare in the mid–nineteenth century), adequate local food supplies were something that the local elite could take for granted. Furthermore, large farmers were more likely than subsistence peasants to be well informed about economic alternatives and to have the capital with which to finance new crops and new techniques. And for these very reasons, they were more sensitive to considerations of profitability, even if its pursuit entailed some risk.

What is unusual about the agricultural history of this *commune* since the middle of the nineteenth century are the abrupt shifts of direction in response to economic opportunities that have taken place, though the particular circumstances of such changes of course cannot always be traced. Take local stock raising, a relatively marginal activity in Saint Cézert, as an illustration: throughout the nineteenth century (at least as late as 1886), villagers kept flocks totaling six hundred to seven hundred sheep, guarded by a communal shepherd. By 1914, their number had declined to about five hundred, yet only a few years later, in 1921, sheep were down to ten. True, wool prices had been going down on the world market, yet what makes the reaction of the Saint Cézert farmers peculiar is that within seven years everyone got rid of practically all their flocks.[13] Another, perhaps related, example: throughout most of the premotorized agricultural era, local farmers got along with seventy to eighty work-oxen and only a handful of cows. Yet suddenly between 1914 and 1921, perhaps in response to the wartime demand for beef and the shortage of farm labor, stock raising took off. Including the usual fifty yokes of work-oxen and a couple of dozen milk cows, Saint Cézert came to boast of a cattle herd of almost six hundred head in 1921, 50 percent more than the maximum bovine head count in Juzet d'Izaut, long a cattle-raising village. Was it a matter of trading sheep for steer? The cattle boom ended as abruptly as it had begun: by 1931, livestock was back down to 104 head, only a couple of dozen more than in 1914.[14] The abruptness, frequency, and unpredictability with which agriculture in Saint Cézert has shifted direction since the middle of the last century lends the community its singular character. As we shall see, the tradition of being antitraditional is still alive and well.

The major phases of Saint Cézert's agricultural development since the early nineteenth century are clear enough, though details are sometimes hazy and the chronology leaves much to be desired. For the first two-thirds of the last century, the community's farmers were chiefly oriented toward grain production, although about 80 hectares not suitable for grain or grass were planted in vineyards. As had been true at Buzet, in the early decades of the 1800s, rye and mixed fields of rye and wheat (what the French called *méteil*) competed with wheat, with wheat triumphant by the 1880s. Crop yields were low. The 1821 official report indicated a ratio of 4.5 hectoliters of grain harvested for every hectoliter sown.[15] This translates into a net wheat or rye crop (after setting aside seed grain for the following year) of 525 kilograms per hectare, almost the same figure as that in Buzet a decade later. According to the local schoolteacher, by the 1870s, Saint Cézert's farmers produced a wheat crop of 400 metric tons in a good year. Apparently by that time, land in *la rivière* was being cropped continuously, while the poorer soil of *la plaine* was left to lie fallow every other year. If we accept the annual consumption figure of 300 kilograms of grain per local inhabitant provided in an 1821 estimate and also allow for seed grain, farmers must have been left with a marketable surplus of 220 tons, worth about Fr 60,000, given average wheat prices in the decade ending in 1872.[16] Wine, which was certainly produced in excess of local consumption, was likely to have provided some additional cash income.

A second phase began in the mid-1870s when the downward slide of wheat prices—a worldwide phenomenon—led Saint Cézert farmers to shift part of their acreage from wheat to the variety of alfalfa known in France as *luzerne*, and, to a lesser extent, to oats. Wheat production fell to half of what it had been, but 400 metric tons of hay were produced and sold. Some of this production of fodder crops probably reflected the substitution of such crops for the traditional fallow, but this does not account for the actual decline in wheat grown. The economic disadvantage of hay as against wheat production suggests that hay may have been no more than a by-product: local farmers may have been growing *luzerne* primarily for seed, a specialty upon which at least one farmer, Masdé, relied as late as the 1970s.[17] Corn, grown wherever the soil retained moisture in the summer, which was mostly in *la rivière*, fed enormous flocks of poultry that Saint Cézert children looked after and that Saint Cézert women sold at the Grenade market and elsewhere. In the meantime phylloxera had practically wiped out local grapevines, though by 1886 some 35 hectares had been replanted, presumably grafted to American rootstocks. This phase of commercial polyculture seems to have lasted to about the turn of the century.

A third phase, which continued into the 1970s, saw Saint Cézert's farmers turn decisively from polyculture resting on wheat, *luzerne*, and oats to wine production, with wheat running a distant second and barley even further behind. Oral tradition has it that a senator and regional notable, who had created a vast nursery along the Garonne River specializing in American rootstocks, persuaded his friend, the then mayor of Saint Cézert (and a member of a major farm family still prominent today) that phylloxera-resistant vineyards were an opportunity too good to miss. However unsubstantiated, the story is at least plausible. The fact is that even though some adjoining villages had very similar gravel soils, Saint Cézert was alone in becoming a township of vintners. By 1914, the *commune* had 181 hectares of vineyards in production and 31 hectares of recently planted vines.[18] This does not imply that everything but grapes was neglected, if only for the good reason that no more than a quarter of the local land had the combination of sunny exposure and well-drained soil that vineyards required. Except for a few years in the 1960s, when a North African *rapatrié* planted 60 hectares of grapes on the domain of Entenepay that he had purchased, vineyards in the *commune* never occupied more than 215 hectares. Yet some contemporary farmers remain convinced to this day that there was once some mythical era when 500 or even 600 hectares of Saint Cézert soil were covered with vines. Like other myths, this exaggeration may be a way of getting at what had locally come to be regarded as an eternal verity, namely the notion that, as one farmer and municipal councilor put it in 1945, "the cultivation of the vine is the only venture with some potential that has actually met with success here."[19]

What really counted, after all, was the relative income from wine as against income from competing crops like wheat. We happen to have fairly reliable statistics for 1944, a year which, the war notwithstanding, may be considered reasonably typical. In that year there were 211 hectares of vineyards in the *commune*, as against 155 hectares of wheat, the second-most important crop, yet the gross income from wine was nearly eight times that from wheat. In 1944, local farmers also grew 66 hectares of oats, 26 of corn, 3 of dry beans, 5 of potatoes, 69 of artificial meadows, and 61 of natural meadows: this was continued polyculture all right, but it was wine that brought in the money.[20]

The early beginnings of mechanization in Saint Cézert are not distinctive, though new machinery and new tools tended to appear a few years earlier than in Buzet-sur-Tarn and considerably earlier than in Juzet d'Izaut. What must have made the difference was the predominance of solid, business-minded farmers who had the reserves to buy equipment that made sense to them. The two local estates, La Mothe and Entenepay, run by bourgeois absentees who had farm stewards, were not likely to be in the forefront of modernization, for it was difficult to break with an old regional tradition of low investment in such estates. It was the large owner-operators who were more likely to take the lead. By the time of World War I, for example, the big farmers (*les gros*) had generally acquired the harvester-binders that small farmers (*les petits*) were not able to afford until the 1930s. For one thing, *les gros* could afford the two yoke of oxen by which this powerful piece of machinery was normally pulled. When *les petits* got around to buying their harvester-binder, they had to improvise a troika by hitching their one horse (normally used for work in the vineyards) to the yoke of their single pair of oxen.

The agricultural survey of 1929 showed not only the general use of equipment that one might well expect in a *commune* such as Saint Cézert, like the twenty-seven heavy-duty sprayers for grapevines (at a time when Buzet, with only slightly smaller vineyard acreage, had no more than fifteen), but also twelve disk harrows, three mechanical seed-drills, and three farm trucks, equipment that, for all practical purposes, was not seen in Buzet before the 1950s and in the case of trucks not even then.[21] In both communities, the single tractor (Saint Cézert also claimed the ownership of a nineteenth-century type of steam-driven "locomobile") may have been in use either by a local harvesting contractor or, in the case of Saint Cézert, by the La Mothe estate for hauling and powering threshing equipment.

In some cases, the availability of the new equipment, especially of seed-drills, made a considerable difference in crop yields such as that of wheat. In 1945, Bouquet, one of the smaller farmers who undoubtedly continued to broadcast his seed, considered a net yield of 650 kilograms of wheat per hectare as average for the *commune*.[22] Another *petit*, Bissaire, remembered the ratio of seed to crop as having amounted to 1 to 5–7 in the 1930s and 1940s (compared to 1 to 4.5 in 1821) with a net harvest of 600 to 1,000 kilograms per hectare (compared with a net wheat harvest of 525 kilograms in 1821). Yet one of the *gros*, whose farm undoubtedly already employed the seed-drill, insisted that 1,800 kilograms would have been a more realistic approximation.

Part of the discrepancy may have had to do with agricultural technique, but another part may have reflected drastic fluctuations from one year to the next. For instance, while, according to official figures, the 1914 harvest had averaged 1,275 kilograms per hectare, the 1921 average was 600 kilograms.[23] One week of *vent d'autan*, the searing wind from the southeast, at the moment when wheat was flowering could halve the harvest. All that one can say with confidence is that some progress had been made since the early nineteenth century, but that the extent of that progress was unspectacular. This is not surprising: even the big farmers used only trifling amounts of a mixture of superphosphates and lime as fertilizer, though oxen and horse did provide some dung and fodder crops some nitrogen as well. In a sense, even the per hectare figures for these modest crop yields overstated the reality. Much of the arable upland continued to rest every other year, ritually plowed up three times over spring and summer, but remaining unsown. This signified that, in the long run, the effective returns in the highlands—four-fifths of the arable land—were only half as high as the figures that we have examined.

As far as organizing the essential tasks was concerned, methods remained tradition-bound. Take the 100 hectare Masdé farm, which, in contrast to Saint Cézert's two estates with their pretentious chateaux, was centered on a large, ramshackle farmhouse, with the family sandwiched between the stable housing the oxen and the one sheltering the horses. Even though the then reigning Masdé had bought a tractor as early as 1934, as late as the end of World War II, plowing was still done by seven yokes of oxen, cultivating the vineyards by three horses. There were seven, sometimes eight, year-round employees who were lodged in what the present Masdé described ruefully as rural slums, but which were taken for granted at the time.

For each of the plowmen, work proceeded according to a traditional ritual. Each man was in charge of one pair of oxen (*boeufs* in French, hence he was known as a *bouvier*). During plowing season, the workday for the *bouvier* began at 4:00 A.M. and ended at 7:00 P.M. He first fed and watered his oxen, leaving the stable by 5:00 A.M. and plowing until

7:00 A.M. Plowing was done with a reversible Brabant, an efficient plow, as noted earlier, that could be set to a given depth, but with a plowshare only six inches wide, a third of the width of even small tractor plows today. Narrower plowshares meant more time spent plowing in that more furrows had to be plowed. Between 7:00 and 8:00 A.M., the *bouvier* took a one-hour break for breakfast in the field, while his oxen rested or grazed along a hedge or in a ditch. From 8:00 A.M. to 12:00 noon, he plowed once again, then took his pair back to the stable to water and feed. Then the *bouvier* quickly had his lunch at home, so that he would have time to catch forty winks before getting his yoke of oxen from the stable. He plowed again between 1:00 P.M. and 4:00 P.M. At that hour someone was supposed to bring him his afternoon snack, while once again the oxen were given their rest. Plowing resumed at 5:00 P.M. and ended at 7:00 P.M. After taking his animals back to the stable, watering and feeding them, what was left of the day was his own—provided that he kept in mind that he would have to arise no later than 3:30 A.M.

With farms that produced both grain and wine, there was hardly any slack time during the year. Besides plowing for crops and for fallow, there was dung to be spread and fields to be sown, meadows to be cut and hayed, wheat to be harvested and threshed, ears of corn to be hand-picked and shucked, and weeds to be kept down by cultivating—shallow-plowing. The vineyards also had to be sprayed and trimmed all summer before being cut back during the winter months. And during the grape harvest, not only were all local hands kept busy, but for four weeks the village swelled to three times its normal size as migrant workers, peasant families from nearby villages, from the Pyrenees and the mountains of the Auvergne, but some from afar afield as Spain and Portugal, flowed into Saint Cézert for the harvest.

The substitution of tractors for oxen came earlier to Saint Cézert than to either Buzet-sur-Tarn or to Juzet d'Izaut and the process was less protracted. Between 1945 and 1948, agriculture in Saint Cézert became largely motorized. The oxen at the Masdé farm went to the butcher, with the exception of a single pair retained to keep the oldest farmhand happy until he reached retirement in 1950. When he went, so did the last oxen, by which time work animals were rapidly becoming a local anomaly. As elsewhere, the changeover was slower for small farmers. Young Guichot, for example, whose parents owned and farmed 18 hectares—a sizable farm anywhere else in the area at the time, but one that clearly put them among *les petits* in Saint Cézert—started out by plowing with an American jeep, the gearing of which he had changed to increase its power. The Guichots did not buy a regular tractor until several years after Masdé had got rid of his last pair of oxen.

Initially, mechanization brought less obvious socioeconomic change to Saint Cézert than to either Buzet-sur-Tarn or to Juzet d'Izaut. In Buzet, the tractor, because it was a heavy investment that had to be amortized by small peasants whose primary concern had been self-sufficiency, led to the intensification of agriculture by way of dairying and to an extension of acreage farmed by way of renting or buying additional land. In Juzet, where the much later introduction of tractors coincided with land consolidation, both the size of farms and the size of dairy herds tended to increase drastically. Nothing quite like this occurred in Saint Cézert, because the *commune* was already dominated by large farms and because, until the 1960s, very little land came on the market. Buying tractors to replace oxen did not force someone like Masdé to expand his land holdings, because 100 hectares, of which 33 were in vineyards, easily justified several tractors to replace the fourteen oxen and three horses. This was true even for farmers who were not of Masdé's stature, like Dupouteux, of whose 30 hectares one-half were in grapevines. Even a farm of this size, which supported two full-time hired men who worked alongside father and son, was sufficiently productive to bear the new overhead.

For the same reason, buying tractors did not force farmers like Masdé or Dupouteux into serious dairying either. In any case, such a transition would have been much less natural in Saint Cézert than it was in Buzet, even before the arrival of the Aveyronnais dairymen. When farmers in Buzet had bought their first tractor, they were generally left with a couple of now unemployed work-cows that had always produced yearly calves and that could now become the nucleus of a stock-raising or dairying operation. In contrast to Buzet, most farms in Saint Cézert had not relied on work-cows but on oxen, which, once they lost out to tractors, were fit only for slaughtering. Even so, for a number of years stretching into the 1970s, a fair number of farmers did keep a few milk cows, usually between three and six, that were milked by hand. They provided a small, but steady, year-round cash income and, presumably, continued to supply the manure formerly produced by work oxen.

Although direct evidence is unobtainable, in Saint Cézert the more abrupt substitution of machines for livestock was bound to induce local farmers to use chemical fertilizers much more intensively than ever before. On the larger farms, the end of oxen and, a little later, of horses meant the end of available animal manure. Work animals, it is true, had never provided enough dung to go around, a widely recognized fact that accounts for the widespread practice of fallowing in alternate years. Yet land to be sown *was* regularly manured. The presence of work animals and their need for winter fodder had also induced local farmers to plant nearly an eighth of the township in leguminous grasses like alfalfa that renewed the soil with their nitrogen.[24] When oxen and horses went, so did these natural and artificial meadows. The only ready solution available to Saint Cézert farmers was heavy reliance on the much wider range of sophisticated fertilizers that were becoming available, particularly nitrogen and mineral-rich slag (*scories*) from iron foundries. Though local wheat would never overcome some of the natural handicaps of soil, wind, and weather, artifical fertilizer systematically applied did double and triple grain yields and, from one year to the next, doubled the amount of available arable upland by making fallowing obsolete.

Where dairying operations of any size did crop up in Saint Cézert, they had less to do with the coming of the tractors than with individual farmers' prior experience and habits. Bouquet, who on 17 hectares not only supported his family, but sheltered and fed half a dozen Jews in hiding during the war years, did so by keeping a dairy herd. The butter that he made went to a city baker, in return for which Bouquet and his charges were supplied with bread with no questions and no ration coupons asked. In his case, anyway, dairying began long before tractors, although he kept a herd until his son took over in the 1970s. Yet he was simply doing what he knew best, having been a dairyman on the plateau of the *Montagne noire* before a bovine epidemic forced him to relocate in Saint Cézert in the early 1940s. Another dairyman, Franchini, who came to the township as a sharecropper and whose son enjoys the reluctant distinction of being the last sharecropper in the *commune*, was similarly carrying on the dairying tradition of the region of Northern Italy from which he hailed. In the case of a more recent arrival, Salut, who purchased the worst chunk of Entenepay in 1978, dairying was dictated by the fact that the land was not good for much else, but then he also came from a part of the neighboring *département* of Gers where dairy cows had been kept as a matter of course. All of these farmers were outsiders. Most local people, even *les petits* like Guichot, Gautier, or the brothers Bissaire, were reluctant to take on cattle once they had got rid of their work-animals.

By the 1950s, when the new high-investment agriculture was beginning to pinch, those of the small farmers who had enough family manpower increased their income by growing tobacco, a crop tightly governed by the French national tobacco monopoly. In the process, in the early 1960s, they also pioneered in irrigated farming by pumping water from the brook to help their precious crop along. Gautier, for instance, who now farms 40 hectares, at that time owned only 5 hectares, which had to support not only himself, his wife, and his children, but both his parents and parents-in-law as well. With an immense amount of hand labor it was possible to make a living on less than 1 hectare. He remembered delivering as much as 4,500 kilograms of dried tobacco to the Toulouse state tobacco warehouse—an amount that (in 1982) would have fetched Fr 180,000 at the official top price.

Even though 1 ton of tobacco had become the equivalent of one year's legal minimum wage, Gautier and the other tobacco growers hardly looked back on this phase of their life with nostalgia. As his neighbor, Guichot, explained, growing tobacco was not only endless, painstaking work, but unending frustration in dealing with the state monopoly as well. They had once tended 12,000–14,000 plants on half a hectare. Harvesting had to be done on hands and knees, one leaf at a time and starting at the bottom of the plant. The complete harvest required returning three times to every plant, as only mature leaves could be selected. Once the leaves were dried in the shed that Guichot had built from his own poplars, the tobacco was taken to Toulouse for sale. What had first been a bonanza, increasingly became a headache. As the supply of regionally grown tobacco grew, buyers for the state monopoly became increasingly choosy: his leaves, Guichot would be told, did not burn well enough; they were too high in humidity, too light in color, or too green altogether. The official buyers discovered a gamut of good reasons for never paying top price. After a dozen years of this, Guichot and other small planters in Saint Cézert were sick of it and quit, but only after the profits from tobacco growing had eased their transition to high-investment farming. In Gautier's case, it also appeared to have given him a leg up in providing the down payment for additional acreage once local farmland came on the market. Tobacco growing had been the underpinnings for an up-to-date farm operation, although excessive debt left it economically vulnerable.

Between 1965 and 1978, roughly one-third of the land in Saint Cézert changed hands, as radical a change in land ownership as any recorded. Some of this upheaval was nothing more than the chance result of inheritance. In the earliest instance, one real estate transaction involved the 40-hectare estate of Sapène, which the schoolmaster and nineteenth-century chronicler of Saint Cézert had devoted his life to accumulating. When Sapène's daughter and son-in-law died, the granddaughters, none of whom had remained in the village, settled their inheritance by selling out. The other two instances were typical examples of a characteristic byproduct of French agricultural modernization, namely, the failure of large estates owned by men who were not themselves working farmers to adapt to new conditions.

This was clearly the case of La Mothe, once the property of the aristocratic lords of Saint Cézert, whose twentieth-century bourgeois absentee owners could not cope with an agriculture requiring heavy capital investment and facing rapidly rising wage costs. The liquidation of a second major estate of 127 hectares, straddling Saint Cézert and adjoining Burgaud, called Entenepay, took place in two installments. The original owners had been forced to sell by the late 1950s under conditions analogous to that of La Mothe's proprietors. The buyer was a *pied-noir*, a French expatriate from Algeria, who had been a large-scale landowner and gentleman farmer in North Africa. He proceeded to borrow, invest, and improve on a scale never seen before in the area: no one had ever replanted over 60 hectares of vineyards in one fell swoop, yet the lavish expenditure on vineyards and equipment, on renovating and modernizing the estate's chateau, on hiring and supervising the numerous hired hands, did not stave off his creditors. In time, the *pied-noir*, who was also beset by family tragedies, went bankrupt, though with greater panache than his absentee predecessor.

To make the most of what they saw as a series of unprecedented opportunities for enlarging their operations, local farmers called upon a uniquely French agricultural agency, invariably known by its acronym of SAFER, to act as honest broker and arbiter in these real estate transactions. SAFER stood for Société d'aménagement foncier et d'établissement rural, loosely translated as Land Improvement and Rural Settlement Corporation. About half of the nearly 300 hectares involved in these transfers ended up enlarging the property of farmers in Saint Cézert and of adjoining Aucamville owners. The other half was acquired by three outsiders, Salut and the brothers Duparc, the first new farmers to settle in the township since 1941, when Bouquet senior had bought what was then a farm of 11 hectares. In turn, the Duparc brothers, intent on making the most of local possibilities for profitable grain farming, unknowingly helped nudge the community's winegrowers in the same direction.

The SAFERs, originating in the wave of agricultural legislation of the early 1960s, were intended to cater to the needs of the farming community by regulating what was seen as a chaotic and speculative land market. Set up as regional nonprofit corporations responsive to farm organizations like the chambers of agriculture, farmers' unions, and cooperatives, the SAFER's task was to see that agricultural land ended up with the most deserving farmers—farmers that needed to enlarge in order to remain on the land and young farmers in search of a farm. They did not have the power to expropriate land, but they did have the authority, though not always the financial resources, to preempt a private buyer by offering the market price. They could then make whatever improvements the land or the buildings might need and look for the most equitable way of disposing of the property. The SAFER could not hang on to this real estate, but had to sell it as expeditiously as possible. Even though they were not designed to make a profit, neither were they charitable organizations: their working capital had originally come from state-sponsored agencies like the Crédit Agricole, the gigantic cooperative farmers' bank. Therefore in order to continue operating, they did have to price the land that they resold so as to break even and safeguard their working capital. They were and are in the curious position of being real estate brokers with a social mission. In everyday life, they are also susceptible to political pressures: politically it is usually easier to divide a property among all interested neighbors than to make unpopular value judgments as to who *really* needed more land and who did not.[25]

This is essentially what happened to the Sapène property, to the bulk of Entenepay, the *pied-noir's* estate, and to a fraction of La Mothe. Small farmers like Bouquet, Gautier, and Dulac were able to enlarge, but so were farmers like Rival and Masdé who were already among the most prosperous land owners in the community. One section of Entenepay happened to be the worst land in Saint Cézert: locally no one wanted any part of it. The SAFER sold these 22 hectares to an outsider, Salut, who was shrewd enough to know what he was getting, but who could not afford the price of more desirable land: it was he who started a dairy herd as the one way of making such land pay. The estate of La Mothe was a different story. It consisted of a cohesive block of 150 hectares of the most fertile alluvial soil in the northeast corner of the township, with another 100 or more hectares spilling over into adjoining Aucamville. Here the SAFER seems to have made a more deliberate effort to install would-be farmers in search of a farm. One hundred and sixteen hectares of the best grain-growing land in Saint Cézert were sold to the two brothers Duparc.

In a previous chapter we encountered a number of farmers whom the French confusingly insist on calling *migrants*, who are not migrants in the Anglo-American sense of the word, but outsiders. At Buzet-sur-Tarn the dairy farmers from Aveyron, French émigrés from North Africa, and North Italian ex-sharecroppers all played a notable role in the modernization of local agriculture. The *migrants* settled by the SAFER in Saint Cézert represented somewhat different types, making their own distinctive impact on the community to which they came.

Salut hailed from Gers, the *département* that borders the Haute-Garonne on the west, from an area located no more than 50 miles from Saint Cézert. Both socially and geographically, his was a family of perennial *migrants*. His grandfather, a landless farm laborer, had, by buying a field at a time, succeeded in creating his own farm of 9 hectares in his home village. When local resentment against the upstart precluded the grandfather from acquiring more land there, the family pulled up stakes in the early 1950s to buy an abandoned farm of 27 hectares in another canton, a farm which prospered despite their new neighbors' predictions of doom and gloom. However, family conflict arose a number of years later when the grandson wanted to modernize and expand and found himself thwarted. In frustration, the young man left in a huff. After learning the trade, for twenty years Salut worked as an artisan mason in a number of rural communities in Gers and Haute-Garonne without ever forgetting that what he really wanted to do was to farm. The mediocre acreage he had bought in Saint Cézert was part of a larger, cooperative farm operation in conjunction with his married son and his son-in-law. The rest of their GAEC's (Groupement agricole d'économie en commun) combined 80 hectares comprised rented land scattered in three other nearby *communes*. Only irrigation, a keen alertness to new crops (they were alone in Saint Cézert to plant *feverolles* [animal feed beans], and doing very well with them), and a rapidly improving and expanding dairy herd allowed three families to survive on such limited acreage.

The newcomers' links to the local community were tenuous. Salut, who had been at Saint Cézert just three years when he spoke to me, was on casual terms with two of his neighbors, Gautier and Masdé, with whom he had on occasion traded equipment for a day, but otherwise he went out of his way to avoid involvement with Saint Cézert's farmers. While working as a rural mason in a number of villages, Salut had suffered from local factionalism and wanted no part of it now that he had settled down. Besides, his efforts to build up his dairy herd were unlikely to interest local people and most of his farming was done outside of Saint Cézert anyhow.

The history of the Duparc brothers was very different. Their father, now semiretired and living with them, had been a prosperous tenant farmer in the *département* of Pas-de-Calais, along the English Channel in northeastern France. When, at the beginning of the 1950s, the Duparcs decided to mechanize, which, in their case, meant shifting from horses to tractors, they began to look around for a larger farm and found themselves priced out of the market because of competition from both local would-be purchasers and Belgians. They therefore shifted their search to the area south of Paris, where land prices were cheaper, ending up in 1953 with a 76-hectare farm on rolling terrain not far from Blois, in the Loire chateaux country, where they combined dairying, hog raising, and grain farming. Their farm was just south of the Beauce region, the richest grain plain in all of France, and, indeed, the one Duparc daughter married a Beauceron farmer. Through these new relatives, the Duparc sons learned to appreciate the advantages of grain farming over the "slavery" of keeping cows and pigs, but their farm was too small and too hilly for anything else.

Consequently, the two sons determined to find this larger, level farm of their dreams and to concentrate on growing grain. When they came to look for their ideal farm, they soon became aware that they had entered into unequal competition with rich Beauceron grain growers, who were setting up their younger sons on farms bought in the Loire valley. The Duparcs, father and two sons, began to look south of the Loire, spending two entire years responding to classified ads and inquiring at notaries' offices, for in France, notaries, as obligatory middlemen in real estate transactions, are the most knowledgeable people when it comes to the local land market. Only after having cut the umbilical cord by selling their Loire farm did they drop in at the SAFER offices on the outskirts of Toulouse that referred them to the La Mothe property, which was smaller by 25 hectares than what they had been looking for. The fact that the land was excellent by southern French standards—although no agricultural land south of the Loire compared with that of the Parisian basin or their native northeast—and that an irrigation network reached within 1 kilometer of the property proved decisive.

They harvested their first crops in Saint Cézert in 1974, but needed two or three years to adapt to unfamiliar soil and climate by watching and talking to Saint Cézert farmers. On the other hand, they brought with them sophisticated equipment for soil preparation unfamiliar to native farmers, who went out and bought themselves the rotative harrows and reversible plows that the Duparcs had demonstrated. Unlike Salut, the Duparc brothers made a genuine effort to become part of the village community. They were respected and, up to a point, accepted, yet they remained strangers who spoke careful standard French among southerners who, even when they weren't conversing in occitan, rolled their rs and flattened their diphthongs in that most imitable regional manner that for generations has been staple fare for French stand-up comedians.

In the case of Saint Cézert, the acquisition of the Sapène property by the SAFER and the decision to sell it to the half dozen farmers who were interested in enlarging led directly to the *commune*'s demand for land consolidation.[26] By way of *remembrement*, the Sapène land could, it was felt, be reallocated more intelligently, without forcing some buyers to disperse their agricultural efforts even further by having to work Sapène fields that were too far from their own farm. There is some poetic justice to this, for in his report of 1886, at a time when there was no viable mechanism for land consolidation in France, Sapène himself had been unique among other schoolmasters in devoting a couple of impassioned pages to his township's need to rationalize its inconvenient field system.[27] To be sure, in the 1970s, there had been some prior psychological preparation: the local agricultural study group, the CETA (Centres d'études techniques agricoles), had long advocated such rationalization and some of its members had indeed exchanged fields to get rid of the nuisance of enclaves. These *échanges à l'amiable* (friendly exchanges) were encouraged by the national authorities by exempting such transfers from the otherwise ubiquitous real estate transfer tax and by providing professional, salaried mediators.[28]

There were three elements that made land consolidation a very different process at Saint Cézert as compared to other agricultural communities. One of these has already been pointed out, namely that the catalyst was the allocation among a number of farmers of an estate that was to be dismembered. A second had to do with the predominance of wine growing in the township. Within the *département* of the Haute-Garonne, among *communes* where wine growing predominated, this one was the first to launch a *remembrement*, although a few years later the Fronton area, a larger and better known township of vintners, followed suit. It was widely assumed in the mid-1960s when Saint Cézert was drawn into the consolidation process, that because they embodied so much greater investment in labor, money, and ego than fields and meadows, vineyards would be much more difficult to reallocate. The problem of classifying such land was also deemed infinitely more intricate, as the variables that determined the quality of a vineyard were so much more numerous.

A third unique factor was the role of Saint Cézert's mayor, Thomas Dulac. Normally, when a *commune* committed itself to *remembrement*, unless some neighboring village had already experienced the process, the mayor had no more clear ideas as to what to expect than did his local constituents. This meant that when conflicts arose in the course of land consolidation as they usually did, he would be at a gross disadvantage in dealing with officials and experts who had both superior authority and experience on their side. There were a number of typical postures that a village mayor might adopt under the circumstances: side with those of his constituents who were most alienated; try to persuade the discontented to be good sports and go along; withdraw and pretend that the whole thing was out of his jurisdiction; act as buffer or go-between interposing himself between the authorities and its minions and his own constituents. By and large, his chances of being effective were slim: as an amateur among professionals, a village mayor typically lost control of the land consolidation process and was often voted out of office for his troubles.

By contrast, the mayor of Saint Cézert was a most atypical case. Although a winegrower by background, he had first been elected to the presidency of the departmental federation of farmers' unions, and later appointed to an important executive post as legal councillor with the Chambre d'Agriculture in Toulouse. In that capacity he sat as agricultural representative on the departmental *remembrement* commission that heard appeals from complainants in various land consolidation cases throughout the Haute-Garonne. He therefore approached *remembre-*

ment in Saint Cézert as someone thoroughly familiar with its legal framework and its practical problems. In short, he could match the surveying staff and the civil engineers of the DDA (Direction départementale de l'Agriculture) in expertise and he outclassed them in clout and political skill.

Once *remembrement* was launched in his own community, he offered the authorities seemingly complete cooperation. The respect he enjoyed among his constituents was such that he was able to settle disputes as they came up, taking surveyors and engineers off the hook. The latter may not have been fully aware of the extent to which there was an unspoken quid pro quo. Unintimidated by legal niceties and bureaucratic procedures, Dulac had every intention of keeping control of the direction of *remembrement* in his village. He was determined that it meet his goals and those of his constituents as he perceived them.

As a result, land consolidation proceeded with great dispatch in Saint Cézert once the cumbersome machine was officially set in motion with the usual two-year delay between the official local request and the first meeting of the communal *remembrement* commission. Headed, as is normally the case, by a judge and including representatives of the authorities, the commission was skillfully manipulated by the mayor, who saw to it that the local farmers selected for membership would play a determining role in shaping its plans. Between the first meeting of the commission in December 1968 and the reallocation of the newly drawn fields in November 1971, less than three years elapsed. Land consolidation in Saint Cézert was carried through at a pace that few other communities in the region could match.[29]

More important from Dulac's viewpoint was the fact that he managed to remain fully in charge. The consolidation of the vineyards was a case in point. Ten years after the events, the agricultural civil engineer in charge at Saint Cézert glowingly recalled land consolidation there as a splendid pioneer venture, because for the first time in the Haute-Garonne a significant acreage of grapevines had been consolidated. He himself had drawn up an elaborate classification scheme, which took into account such variables as the age of the vines, the grape variety, whether or not the rows permitted passage of a tractor, the degree of sunny exposure, as well as the suitability of the soil. It was true that, in contrast to Buzet-sur-Tarn, for instance, the vineyards had been officially included in the perimeter to be consolidated.

Yet when local vintners discovered that, even with the most sophisticated classification system conceivable, forced exchanges were highly unpopular, they simply changed the rules of the game. No one would be forced to trade his own vineyard for a neighbor's, though exchanges that seemed feasible and convenient were encouraged. In fact, relatively little wine-growing land changed hands, yet this sabotage of the recognized legal procedure—any land officially included within the consolidation perimeter was bound to be consolidated with or without the consent of its owners—was sufficiently subtle, to the point that the government engineer never caught on to the fact that he had been manipulated. The surveyor, who could not be hoodwinked so easily, was simply worn down by passive resistance until he came up with a field plan conservative enough to be acceptable to the local farming community. In retrospect, by preventing the consolidation of the vineyards, Saint Cézert farmers may have outsmarted themselves, but in the late 1960s no one could foresee the possibility of a future when wine growing might outlive its time.

Under their mayor's leadership, the local farmers also applied the brake to the grander ambitions that normally animate such land consolidation projects. The official, stated philosophy of *remembrement* is to try reallocating the land in such a way as to group each farmer's acreage around his residence, ideally in one single block of landed property. Both because they had vineyards dispersed throughout much of the township and because the quality of local agricultural land varied so widely, Saint Cézert farmers had no use for the official goal. Land ownership was to remain dispersed, farmers felt, but instead of having scores of scattered fields, they wanted them consolidated into larger, more workable, units. Masdé's 100 hectares divided into about sixty fields ended up as eight blocks of land; Dupouteux's 30 hectares in thirty parcels were consolidated into ten larger fields. There were examples where what had been seventeen intermixed strips of land were joined to make a single respectable field of 2 hectares that was allotted to Bouquet. The results were useful, though not spectacular. It was, Dupouteux asserted in retrospect, "*remembrement* with a human face."

Where Dulac was convinced that radical changes were essential, such major changes occurred, even if they were not entirely state-financed as the actual revamping of the field system was. Communal roads were redrawn and asphalted, with every field in the *commune* accessible by at least a gravel road. Moving farm equipment, which had been an axle-breaking nightmare, became quick and painless. When the mayor was reelected after the completion of land consolidation his chief campaign boast had been that he had provided Saint Cézert with roads that had been 80 percent subsidized by the state and the *département*.

Where the mayor had no strong feelings, the officials conducting the operations had free rein. All hedges were razed because this happened to be an obsession of the particular surveyor selected by the DDA at a time when environmental concerns rarely got a hearing. It also turned out to be a costly error in a region where windbreaks are beneficial and perhaps indispensable for agriculture. Maybe because they were too busy wrangling over swapping vineyards (invariably on land that was well-drained anyway), neither the mayor nor the surveyor showed much interest in improving the local acreage devoted to grain growing. In this instance, the community's farmers did not take full advantage of government subsidies. As only a few drainage ditches were dug, many fields remained soggy during spring and late fall. As was true in Buzet-sur-Tarn, this major error of omission would have to be made good within a few years.

The reallocation of land, which the sale of the Sapène, Entenepay, and La Mothe properties had promoted, brought newcomers into the community, allowed major landowners to round out their holdings and, in two or three cases, permitted marginal or submarginal farmers to enlarge their farm to a more viable size. Yet these real estate operations did not really revolutionize the Saint Cézert farming community. The same may be said for land consolidation which, at the outset, was begun to make these land transfers feasible in the first place. Everyone agreed that the results were a success, though, looking back on it, they also noted that the outcome would have been even more positive had *remembrement* been conducted a few years later by which time most of the township's vineyards had disappeared. The new asphalted communal roads were an immense advance over the unimproved and unplanned dirt roads of pre-consolidation days, gravel-covered field access roads did replace boggy tracks, and *remembrement* successfully rid farmers of hard-to-work miniature fields. These changes were important, but they were not central. In a community of vintners only wine and everything affecting wine production is really central.

It is essential to have a feel for the context within which Saint Cézert vintners worked in order to understand the problems they encountered. Since it had launched into grape-growing on a large scale in the early twentieth century, Saint Cézert had all along been producing ordinary table wine, referred to as *vin de consommation courante* in French administrative jargon. This is bulk wine, sold without a label of origin or an indication of the variety of grape from which it has been pressed. In Buzet-sur-Tarn we encountered the peculiar French method that partly determined how much ordinary table wine was worth: such wine, sold by the hectoliter, was paid not according to its taste, bouquet, color, or grape variety, but according to its alcohol content. This did not imply a fixed price. The government's recommended price, *le prix de référence* (in 1982, for instance, Fr 1.85 for each degree of alcohol per hectoliter of wine), was just that, a nonbinding recommendation to the world at large, which echoed medieval notions of the "just price." On the other hand, there was an effective floor under wine prices, namely the price (Fr 1.45

per degree in 1982) at which the government offered to buy wine for distillation into alcohol. The market determined whether Saint Cézert vintners would have to sell their production for distilling or whether they could obtain a better price, usually from Bordeaux dealers for purposes best left unexplored. (In 1982, for example, several sold their wine for Fr 1.65 per degree). What was fixed was the *relative* price based on the percentage of alcohol: whatever the price offered, a wine with a 12 percent alcoholic content would be worth 25 percent more than a 9 percent wine.

As a counterpart to providing minimum price supports for wine, the government, through its appropriate bureau of the Ministry of Agriculture, the Institut des vins de consommation courante, exercises considerable control over wine growers producing ordinary table wine. It controls the total acreage of commercial vineyards and, as we observed in Buzet, under its aegis the right to replant has become a commodity to be bought and sold. It prohibits the planting of some varieties of wine grapes and its inspectors can set deadlines for forcing farmers to pull out such offending varieties. This has been particularly galling to vintners, since some of these grape varieties, mostly American-French hybrids, were originally recommended for planting by the authorities themselves, particularly after the great freeze of 1956. The institute also keeps a changing list of authorized and recommended grape varieties for the consideration by vintners about to replant. Moreover, some control is exercised over what methods are and are not deemed legal in the actual wine making. Chaptalization, for example, by which sugar is added in order to increase a wine's alcohol content (widespread and legal among producers of such quality wines as red Burgundies), is outlawed for ordinary table wines.

The postwar experience of the vintners of Saint Cézert was extraordinarily complex: technical, social, and economic change interacted in intricate ways that they could not have predicted in the 1950s when the process began. To summarize schematically: vineyards have always demanded an amazing amount of tedious, meticulous, mostly semiskilled, work. The paradox is that even though fantastic technical advances have revolutionized grape growing, both by multiplying grape production per hectare and by all sorts of techniques for substituting chemicals and machinery for manual work, that is to say, capital for labor, vineyards continue to require inordinate amounts of human attention, inordinate, that is, given the shrinking pool and rising wages of farm workers. At the same time, stagnant or declining prices for ordinary wines have made it increasingly difficult to pay wages high enough to retain what help continues to be needed.

The intensification in Saint Cézert's wine production has been the most obvious trend, although it has not been achieved without cost. In

1944 a semiofficial estimate put wine production at 37 hectoliters per hectare, although the 1944 harvest may have been on the low side because of wartime shortages of sulphate for spraying vines.[30] In the decades since about 1950, productivity has been revolutionized. Through the planting of new and more productive varieties, but more particularly, by the systematic application of tremendous quantities of chemical fertilizers, by the beginning of the 1980s Guichot, who was noted for lavishing both care and chemicals on his vines, was obtaining 150 hectoliters per hectare and wine yields well above 100 were becoming the average. A production of 100 to 150 hectoliters of wine amounted to a tripling and quadrupling of the yields of 1944. A corollary to this increased production was the construction of new and larger cement fermenting tanks by individual vintners (a dubious blessing to any wine drinker who has ever deplored the unmistakeable "hollow" taste of wine fermented in cement), which went along with a less haphazard approach to the business of wine making itself.

High yields, on the face of it, meant an enhanced income. Yet things were not quite that straightforward, because this type of forced viticulture yields grapes lacking normal sweetness and is consequently liable to end up with wine of unacceptably low alcoholic content: red wines with under 9 percent alcohol are not considered marketable. The higher the yield per hectare, the more likely is this to become a problem. The solution, rumor had it, was that every year before grape harvest time, Saint Cézert vintners accumulated impressive stocks of sugar to help their wine along. These creative wine makers ended up with 11 percent alcohol in their product. In the process they also became reluctant outlaws, given the dim view that the Institut des vins de consommation courante has traditionally taken of chaptalizing.

Increased production as a result of fertilizers and more productive plant varieties was only one aspect of the vineyards' modernization. Labor-saving techniques and equipment were another. The narrow vintner's tractor, capable of fitting between rows of grapevines, took the place of the work-horse soon after regular tractors had displaced oxen. Later, new plantations of vines tended to be more widely spaced to take tractors into account. Major technical advances reducing or eliminating manual labor arrived on the scene gradually, between the late 1950s and the end of the 1970s. The application of herbicides to get rid of grass and weeds replaced shallow plowing of the rows, what vintners quaintly used to call *déchausser la vigne* (to take the vine-stocks' shoes off). Elaborate new pneumatic misting apparatus, spraying four rows at a time and using less than 100 liters of sulphate solution per hectare, replaced the cruder and less economical pressure sprayers in use since the 1920s. In Saint Cézert plants were henceforth trained to grow along horizontally strung wires (though this meant hiring a crew of women each spring to train the grapevines by winding the young shoots along those wires), which opened the way for a machine that mechanically trimmed the grapevine into hedge form. This piece of equipment could take the place of hand trimmers who had been kept busy during spring and summer. Finally, by the late 1970s, grape-harvesting machines had been sufficiently perfected to come into widespread use. Though winegrowers like the fastidious Guichot continued to object to the machines contaminating the grapes with sap and leaf fragments, the new equipment harvested in a day what crews had taken two weeks to pick, a speed that practically freed vintners from the menace of bad weather spoiling the harvest and the wine. The extension after 1970 of the minimum wage to grape pickers, coupled to the increasing insistence on the part of labor inspectors that employers contribute social security payments even for migrants, was inflating grape harvesting expenses. Even though harvest contractors charged a hefty price for bringing in the new machines, most winegrowers agreed that the introduction of this equipment had lowered costs, perhaps by as much as one-half.[31]

The introduction of new machines and new techniques undoubtedly lessened the amount of human labor needed, but it did not mean that vineyards now took care of themselves. There remained an irreducible and still substantial amount of semiskilled hand labor: winegrowers and their hired help continued to spend the winter months in the vineyards on the essential task of cutting back the preceding year's bearing vines.[32] To insist, as the more substantial of the Saint Cézert farmers invariably do, that high wages have done them in, or at least have done them in as winegrowers, seems at first glance absurd. Agricultural wages have always been abysmally low and, by the standards of a prosperous, fully developed, industrial country, they continue to be abysmal. On that abstract level, the notion that this subproletariat of hired hands was squeezing honest winegrowers dry is laughable.

If, on the other hand, we use purely historical criteria—how the real wages of agricultural workers actually compare over time—there is no doubt that in a branch of agriculture traditionally heavily dependent upon cheap labor, labor costs went up spectacularly in the thirty years from the early 1950s to the early 1980s. On a national level, translating agricultural money wages into 1982 francs (with a purchasing power of roughly Fr 6 to one American dollar), the average monthly wage of an agricultural worker rose from Fr 408 in 1951 to Fr 1,434 in 1973.[33] With 1951 as the base year, real wages doubled by 1961, tripled by 1969 and just about quadrupled by 1982. This did not include the costs to the employer of the gradual integration, from the 1960s onward, of farm workers into the agricultural annex to the French social security system or of such

10. An old-fashioned grape harvest

11. Harvesting grapes by machine

fringe benefits as paid vacations and overtime. After 1970, agricultural workers' wages had to meet national minimum wage standards. Though official statistics do not reveal any spectacular jump nationally as a result of extending minimum standards to farm workers, apparently in Saint Cézert wages did rise sharply, indicating that they had been well below the national average. Generally thereafter, the minimum wage, known in France as the SMIC (*salaire minimum interprofessionnel de croissance*), kept well ahead of inflation, even before the Socialist government in 1981 fulfilled one of its campaign pledges by increasing the income of those at the bottom of the economic scale through a boost in the SMIC. Earlier, in 1972, agricultural laborers had for the first time been covered by unemployment insurance, with the cost largely passed on to their employers.[34]

The net result was that between labor-saving machinery and techniques on one hand and rising labor costs on the other, Saint Cézert winegrowers drastically reduced the number of full-time workers they employed. There were thirty-nine year-round salaried workers in 1954, twenty in 1970, and six on only four farms in 1979, at least two of whom worked for farmers who were disabled for reasons of old age or poor health.[35] The example of Dupouteux, a member of Saint Cézert's old families, is fairly typical. Half of his 30 hectares were in vineyards and the farm employed two year-round hired hands when he took over in the 1950s. About 1970, Dupouteux let one of his workers go when the extension of the SMIC to farmworkers raised salary costs 30 percent, dismissing the second one in the mid-1970s. One of his former workers, now retired, but still living rent-free in a house Dupouteux owns, helped intermittently with the winter work of cutting back the grapevines. Rival, who farmed 70 hectares, of which over 30 had been in vineyards at one time, had three salaried employees. By 1983 he was working alone, though he had to hire help for occasional tasks.

One aspect of the squeeze also had to do with the trend of wine prices. Whether or not Saint Cézert vintners were right in putting the primary blame on Italian competition—after all, the Italians could hardly be faulted for the declining consumption of all table wine by ordinary Frenchmen and Frenchwomen—in real terms, and particularly by the 1970s, wine prices did tend to stagnate or actually decline. At best, they rose less rapidly than the minimum wage, the general price level, and even the notoriously laggard price of wheat.[36] As long as wine production per hectare was continuing to go up year after year, the price per hectoliter was not so crucial: a vintner with 15 hectares in grapes, who could count on producing and selling 150 additional hectoliters of wine in a given year, was not going to worry overly as to whether the price went up by Fr 2 or Fr 4 per hectoliter. By the 1970s the intensification of grape production was nearing what appeared to be its natural ceiling, a fact to which the routine reliance on illegally added sugar to the wine-must testified. At that point, if anything, production tended to go down, as the inspectors from the Institut des vins de consommation courante demanded that winegrowers pull out hybrids that had been outlawed because their quality did not match the spectacular yields that had made them popular in the first place. These prohibited grapevines were frequently replaced by more "desirable" varieties that also turned out to be much less productive.

Not all winegrowers felt the squeeze. Ironically, it was the big farmers with sizable vineyard acreage who felt cornered. They simply could not get rid of all their workers, full- or part-time. Whether they harvested with a crew or, after 1980, with a machine brought in by a harvesting contractor, their expenses were considerable and rising, while their revenues were pinched. Yet growing grapes intensively and selling wine on a small scale could still be carried on with profit. While the bigger farmers were hurting—or at least complaining that they were—Guichot, who farmed only 20 hectares, was doing sufficiently well that he could afford to give up growing tobacco, an obnoxious crop, as we have seen. He and his wife normally tended 6 hectares of vineyards on which they obtained the best yields in the township. Both of them did all the work that needed to be done, from cutting back the vines to harvesting the grapes themselves. In 1982, they claimed to net Fr 17,000–18,000 per hectare from their grapes. In this particular set of circumstances, small may not have been beautiful, but it was surprisingly profitable.

Did Saint Cézert have any alternative to continuing its production of ordinary table wine and to working within this system? One such alternative that we encountered in Buzet involved making wine from selected recommended varieties, such as Syrah, Cabernet, Gamay, Jurançon, and the like. Provided the resulting wine reached the requisite minimum of 9.5 percent alcohol, it could be sold as "regional wine" (vin de pays), labeled as to grape variety and regional origin, which might fetch two or three times the price of bulk table wine. The great drawback of this tactic was that these noble grapes were generally less productive than the plebeian varieties and the higher prices might be wholly or partly offset by lower production per hectare. Indeed, in Saint Cézert some small acreage of the recommended varieties had been planted in the post-1956 period, but no attempt was made to market these as regional varietals. Local vintners offered no explanation for this anomaly, except to mutter lamely that it was too much trouble. A likelier explanation is that they were not eager to attract the attention of wine inspectors with palates sensitive to added sugar.

The planting of noble varieties may originally have been intended as the opening gun for a second alternative, namely the long, drawn-out, quasi-judicial, but really political, process by which a wine-growing region seeks official recognition of the superior quality of its wines from the Ministry of Agriculture. It is a process that resembles nothing so much as that of canonization within the Catholic church. In both cases there are two levels of sanctity, the label VDQS (*Vin délimité de qualité supérieure*, that is, wine of superior quality from a specified region) being the equivalent of papal beatification, and the label *Appellation d'origine contrôlée* (AOC) (Named origin guaranteed) corresponding to canonization, properly speaking. These official seals of approval reassure consumers, thereby automatically assuring a higher price in the marketplace, but they also imply increasingly severe regulation of the conditions under which such wines are grown and produced. Arduous and time-consuming as this process of gaining official recognition may be, it is not hopeless. Since the end of World War II, the region around Fronton and Villaudric, on the other side of the Garonne River but no more than 15 miles from Saint Cézert as the crow flies, has succeeded in rising from anonymous obscurity, with a VDQS label as a stepping stone, to an *appellation contrôlée* under the name of *Côtes de Fronton*.

There were a number of reasons why this avenue never really opened up for Saint Cézert. Typically, the initial "miracle" that made launching such a campaign propitious was an already extant regional wine made with a particular grape or blend of grapes that had recognizable distinction. As far as the suitability of its soil and climate was concerned, Saint Cézert was probably no worse than Fronton across the river, yet its best wines had nothing locally distinctive about them, neither in the grape varieties utilized, nor in their overall character or taste. Furthermore, in contrast to the several thousand hectares of Fronton vineyards, Saint Cézert was a little island of 200 plus hectares of grapevines in what was otherwise a grain-growing area. Normally, the minimum acreage considered eligible for official recognition would have had to be in the 600–800 hectare range. Finally, until the early or mid-1970s Saint Cézert's vintners were quite content to go on producing ordinary table wines. They were not even convinced that their colleagues in Fronton, whose wine production per hectare was restricted to one-fourth or one-fifth of their own, were necessarily better off. And by the time Saint Cézert wines were in real trouble, the average age of the vintners was such that to embark on this hazardous quest for earthly glory and VDQS status was out of the question.

Yet by the late 1970s most Saint Cézert vintners, who, like Guichot, owned too much vineyard acreage to manage, increasingly found themselves in an economic bind. Increasingly they also became aware that there was a viable alternative to wine, namely irrigated agriculture and especially irrigated corn growing. As had been the case at Buzet-sur-Tarn, small-scale irrigation in Saint Cézert was nothing new: Guichot had, after all, irrigated his little tobacco plantation as early as the 1960s by pumping water from the local brook, yet this sort of individual effort was necessarily limited: the Margesteau was liable to dry up when its waters were most needed. Thanks to a convenient local well, Bouquet senior had, for a number of years, been able to insure the hay crop he needed for his dairy cows by irrigating some of his meadowland.

Such individual enterprise notwithstanding, in the absence of an available major river like the Tarn, large-scale irrigation in the area only became a practical proposition through the activities of a powerful, non-profit company, what the French call *une compagnie mixte*, combining features of a cooperative with that of a private corporation. This Compagnie d'aménagement des Côteaux de Gascogne (Improvement Company of the Gascon Hills) had a publicly recognized mission to promote irrigation in this region comprising parts of several *départements*. Its earliest nearby irrigation project, in Merville, dated back to the mid-1960s. Adjoining Saint Cézert on the northeast, Aucamville had been linked to the Côteaux de Gascogne's irrigation network a few years later. Saint Cézert farmers were well aware of what was going on in neighboring communities like Aucamville: it did not take long for the word to spread that irrigated corn had turned out to be a smashing success.

By the mid-1970s Saint Cézert farmers began to talk about supplementing or even replacing wine with irrigated corn. Farmers like the Duparc brothers who had bought land in Saint Cézert in order to engage in grain farming were committed to obtaining irrigation from the time they arrived in 1974. So was Bouquet junior, who had left a draftsman's job in Toulouse to take over his father's farm on the condition that he would be free to run it without having to fret over dairy cows or vineyards. Their problem was that nowadays the Côteaux de Gascogne refused to deal with isolated individuals, because it was too much bother. With a rising regional demand for irrigation systems, they insisted that in any one locality there be a critical mass of farmers interested in irrigation. There was also the problem that the Aucamville irrigation network was not infinitely extensible. If ever the central and western parts of Saint Cézert township were to be irrigated, it would have to be through some means other than tying into the Aucamville system.

For those Saint Cézert farmers who were increasingly unhappy with their income from wine, the problem of moving from one year to the next to a completely new agricultural system seemed forbidding. It was one thing to pull out a few hectares of grapevines that the authorities had outlawed or to abandon a handful of overage vineyards for the sake of a more rational field system. It was much more disruptive to uproot what had been both the major investment and the chief livelihood of Saint Cézert for three or four generations. There was also considerable risk involved: it was by no means certain that plowed up vineyards could be planted to corn just like that, given that their soils had for decades been saturated with sulphate and copper sprays. And last, but not least, there was the financial cost of the conversion, of abandoning investments of several tens of thousands of francs per hectare in vineyards in favor of a new and even more onerous investment. The Côteaux de Gascogne had made it clear to local farmers that drainage was an indispensable corollary of irrigation. Drainage alone would cost between Fr 20,000 and 30,000 per hectare (by comparison, local agricultural land in the 1970s was selling for Fr 10,000–15,000 per hectare). Drainage costs would be paid off in installments over twenty years; of course, paying on the installment plan would in turn double the ultimate cost by adding interest payments and making special provisions for inflation. The costs of irrigation would be considerably higher than drainage. It was difficult to disentangle the operating costs—the price for getting a given quantity of water to the hydrant in the field—from the rental-purchase installments on the sprinklers and pipes needed to irrigate. In truth, the obstacle for Saint Cézert vintners was as much psychological as it was financial. After all, in Aucamville and elsewhere, farmers also had their loans to repay in yearly installments, yet compared to preirrigation yields, the new bumper crops of irrigated corn easily repaid the outlay. The difference was that at Aucamville there was no traumatic preliminary of having to liquidate a heritage of wine growing.

The force that moved Saint Cézert vintners was the European Economic Community (EEC) and, specifically, FEOGA, the same European agricultural development agency that Buzet's farmers had besieged to underwrite their irrigation project and which, through the so-called development plans, had ultimately played a key role in Buzet-sur-Tarn's turn toward a high-investment, high-yield agriculture. The role of FEOGA in Saint Cézert differed, but if anything, it was even more crucial. The agricultural economic community had long been troubled by some its member nations'—prior to the admission of Spain and Portugal, only Italy and France were deemed culprits—overproduction of table wine.

FEOGA sought to discourage such overproduction by providing financial incentives for the reconversion of the producers of table wine to some other kind of farming. This took the form of offering for a period of seven years a reconversion subsidy to wine growers (which vintners promptly dubbed *les primes d'arrachage*, the pulling-out subsidies) to get rid of their grapevines and to surrender their right to replant. There were, to be sure, some qualifications. Very recent and very ancient vineyards were not fully covered. Grape varieties that the authorities had declared noxious and illegal on the grounds of producing poor quality or unwholesome wines had to be pulled out without any compensation whatever. Ironically, "noble" varieties like Cabernet, Pinot, or Syrah, the planting of which the authorities had recently encouraged, were also exempt from subsidies for their destruction: neither vintners whose grapes were too bad nor those whose grapes were too good were going to cash in. Only ordinary table wines were in oversupply. The idea of such subsidies was old hat. On and off, they had existed for decades, sponsored by successive French governments. What was novel about the FEOGA reconversion subsidies were the unprecedented amounts involved. We already saw that at Buzet-sur-Tarn, the Common Market subsidies helped close the book on local wine production that was already on its final chapter. In Saint Cézert the subsidies made a more decisive difference because wine growing had been on a so much larger scale there. For vineyard land that fully met FEOGA's specifications, Saint Cézert winegrowers could collect Fr 29,000–30,000 per hectare, two or three times the market price of their land. For a vintner like Masdé, who plowed up well over 30 hectares of grapevines, the reconversion subsidies offered close to a million francs. It was an offer that few felt they could refuse.

In less than three years, Saint Cézert underwent a profound, though perfectly peaceful, revolution. Between the winters of 1978 and of 1981, the *commune*'s vintners pulled out between 100 and 150 hectares of grapevines, leaving only some 40 to 60 hectares of vineyards by 1982. By 1982, every farmer in the community was irrigating, with all but four farmers deriving most of their income from irrigated corn and sunflowers. Speedy as this transformation was, it would have been even faster, had it not been for practical problems that could not be solved in a single year. The Côteaux de Gascogne's extension of the Aucamville irrigation network was limited to the northeastern section of the township of Saint Cézert. Farmers with land there simply signed individual contracts obliging themselves to buy a given quantity of water from the company for at least a period of ten years and acquiring the needed accessories of piping and sprinklers through a rental-purchase plan covering the same period. In 1979 they began to irrigate. It was the Côteaux de Gascogne who served as general contractor for the drainage system installed piecemeal in the course of succeeding winters. The work was about to reach completion during the winter of 1982–1983.

Farmers with land in the south, center, and west of the *commune* chose a variant of the Buzet system by forming a legally recognized irrigation (and drainage) association and hiring the Côteaux de Gascogne to build them a central lake filled by pumping water from the brook during the winter months and doing so at night to take advantage of lower electric rates. Getting this project under way took considerable additional time, so that this second system, a collectively owned and run irrigation lake, did not go into operation until 1982. In fact, the lake was completed too late in the winter to be filled up, so that the irrigation association ended up having to buy emergency supplies of water from the Côteaux de Gascogne to cope with that year's drought. Because the association was heavily subsidized by the departmental administration, the cost of water was one-half of that of the Aucamville–Saint Cézert network owned by the improvement company.

The experience of Saint Cézert farmers with irrigated agriculture paralleled that of Buzet-sur-Tarn, though the fact that at Saint Cézert the Compagnie d'aménagement des Côteaux de Gascogne had insisted on immediate drainage made irrigated farming less hazardous to the long-term survival of the land. Because there were few ancient vineyards in Saint Cézert and because vineyards had been heavily fertilized for a couple of decades, the conversion to corn posed few problems. Typically, Saint Cézert farmers, like those in Buzet, tried to irrigate one-half of their land, switching the irrigated half from one year to the next, so that the structure of the soil would not disintegrate under the pounding of irrigation sprinklers. Sunflowers came on the Saint Cézert scene only in 1981, but almost immediately overtook corn as the more profitable crop: sunflowers brought in every bit as much as corn, but expenses were considerably lighter. All but one local farmer chose the cheaper irrigation system that forced them to spend up to four hours a day moving sprinklers during the hot months. Nonetheless, the officially "reconverted" had become instantaneous converts to irrigated farming. Assured margins of some Fr 5,000 per irrigated hectare made for a very decent living, provided a farmer had enough acreage. For someone like Bouquet junior, who had few vines to pull out and who only owned 32 hectares, irrigation was a great improvement over dry farming, but certainly no road to riches. For others whose farming operations were on a larger scale and/ or who had more vineyard acreage to convert, the reconversion subsidies did provide the wherewithal for exotic agricultural experiments, or for powerful new four-wheel-drive tractors or simply for a nest egg that would permit a more comfortable retirement.

As yet, not all farmers abandoned wine growing altogether. Guichot, for one, only pulled out 3 hectares of prohibited hybrids (for which he received no subsidies) and continued to count on making a living from his 20 hectares by relying chiefly on wine, though he also irrigated corn, adding to his income. Dupouteux, who had grown 15 hectares of wine on his 30 hectares, really more than one man could manage without steady help, cut down to 9 hectares, irrigating half of the rest. Not only was he hedging his bets, but he also found pleasure in being a vintner, the only kind of farmer who followed through on what he grew from the raw material—grapes—to the finished product—wine. The mayor, who had his land worked by a hired man, also kept a fair amount of vineyard, but some of his were noble varieties for which he would not have been compensated anyhow. He, like Guichot, owned only 30 hectares and could not afford to rely entirely on irrigated corn and sunflowers. Rival, who farmed 70 hectares by himself—50 by inheritance, 20 by purchase—kept under 10 hectares in vines (one-third of what he had grown prior to irrigation), but weighed pulling them out before the reconversion subsidies expired in 1985.

Enhanced prosperity conjured up its own set of problems. Rival is a case in point. By 1983, he had outgrown technical training sessions led by agricultural extension agents, but he had just come back from something he considered not only useful, but essential—a three-day workshop on tax accounting for farmers. His problem was a common one in Saint Cézert. Rival was on the verge of being forced to go over to another system of calculating his taxes, an alternative that he was dreading. Most farmers pay a set, estimated tax per hectare of land owned, a system referred to as paying *au forfait*. Every year a departmental commission comprised of representatives of the agricultural profession, of the French equivalent of the Internal Revenue Service popularly known as *le fisc*, and of various other administrative agencies meet to decide that year's taxable average farm income per hectare for each of the geographical regions of a given *département*, in this instance, of Haute-Garonne. As taxable farm incomes tend to be set low, so are taxes paid *au forfait*, particularly for efficient farmers with access to irrigation. The hitch was that, as of 1982–1983 if a farmer's gross income reached Fr 500,000 two years in a row he then had to keep accurate accounts and pay taxes not according to some regionally negotiated and entirely fictitious per hectare income, but rather like any French wage earner, on the basis of his real, documented net income. Rival and his wife, who taught high school, were already paying Fr 20,000 in income taxes, even while he was still *au forfait*. Going over *au réel*, that is, paying at the proper individual rate, was going to be a financial disaster, as he saw it. With a net income approaching Fr 350,000, 45 percent of his income would go for taxes. And doom in the form of *le réel* could only be staved off so long, two years, possibly three years at most. This change in his tax status would in turn force him to pull out the remaining grapevines *prior* to his being caught in the vise of the 45 percent tax bracket or else nearly half his subsidy payments would be taxed as income. And there were other, more Byzantine complications concerning tax alternatives: claiming or not claiming tax deductions on the basis of legitimate farm expenses had predictable, yet incalculable consequences: to legally claim such deductions as local real estate taxes and yearly amortization of drainage costs, Rival would have to post an official assessment of the value of his land. Yet such an assessment registered with the tax authorities would in turn entail having to pay capital gains taxes if ever the farm were sold. Increasingly the quandaries of Saint Cézert's more prosperous farmers were assuming a standard upper-class profile.

Another one of the large farmers, Masdé, who had abandoned his entire vineyard for irrigated corn growing, used part of the funds from FEOGA to launch a venture into exotic fruit growing, following in the footsteps of one of his good friends, a pioneer in introducing the kiwi tree of New Zealand to the French southwest. Kiwi orchards required an incredible investment: the land had to be drained five times as intensively as irrigated cornfields; an underground irrigation system had to be installed to spray the trees as protection against frost whenever the temperature threatened to go down to below freezing; a cypress hedge had to be planted to protect the kiwi trees from the wind. Not counting the price of the land, Masdé had invested Fr 130,000 per hectare, over Fr 800,000 on the 2,400 trees planted on 6 hectares. In 1982, around All Souls' Day, he had harvested his first fruit, just 3 or 4 tons of kiwi, but once in full production—20 to 25 tons a hectare were what his friend harvested—kiwi were a gold mine, at least for as long as the Germans' appetite for this odd fruit that looked like a brown lime needing a shave remained insatiable and French producers remained "disciplined" enough to avoid price cutting.

Nonetheless, these well-off, former vintners who fretted about 45 percent tax brackets and who invested in horticultural gold mines worried about the future of their farms. Among prosperous farmers and even some not so prosperous, children were expected to finish the academic track of the *lycée*, to pass their baccalaureate. If they were so inclined, they were encouraged to enter the university, or, if they had what it took, one of the French elite professional schools. One of Masdé's sons was a hydraulic engineer in Paris; another taught, as did his daughter-in-law, in a *lycée* in a country town at the other end of Haute-Garonne; Masdé's daughter had also finished an advanced degree. Masdé, with some hope for her fiancé who was at loose ends and showed a taste for rural living, was in the process of setting him up with a sawmill in Saint Cézert. The young man, he hoped, would be able to make a living, while getting some practical training in agriculture with his prospective father-in-law.

Even if this worked out, such a succession faced almost impossible obstacles, because under such circumstances French inheritance law calls for the equal division of property. If his two sons chose to demand a legal appraisal of the property, the daughter and her husband would never be able to pay off the money equivalent of two-thirds of the property's market value. The problem was not in itself novel. Masdé was well aware that his, as well as most other, farms had been passed down from generation to generation only because siblings who left did not claim their full share from the brother who stayed. What was new were the extraordinary land prices, prices that drainage and irrigation had driven even higher. Masdé's successor might have a debt of well over Fr 3 million to pay off or Fr 6 million if interest payments were figured in.

Masdé's family situation was fairly typical. Rival, for example, had two daughters who were studying to be, respectively, a computer engineer and a dentist. Neither of their boyfriends had the slightest interest in agriculture. Gautier's daughter and her husband, an engineer, both had good jobs in Toulouse with an affiliate of the recently nationalized conglomerate, Thomson. As for Coutian, an old gentleman in his late seventies, he was still managing his sizable farm with hired help, because his children had long ago left for good. Among the dozen or so full-time farmers in the *commune*, only Dupouteux and one of the newcomers, Salut, the dairy farmer, had an assured succession, though there were one or two more with young children and hope.

Saint Cézert's enterprising farmers, who had apparently succeeded at everything, had also succeeded in outsmarting themselves. In the glow of their very success, they had raised their children for success. The young, in turn, learned to define *their* goals in *urban* middle-class terms by looking to more prestigious professional careers, rather than to farming, however profitable. Prosperity had bought higher education for a new generation and the highly educated rarely returned to the countryside. Saint Cézert's yeoman farmers had not been much more successful in training their own replacements than had the poverty-stricken mountain stockmen and dairymen of Juzet d'Izaut. Despite its tradition of adaptability, Saint Cézert was, ironically, being sapped by the very national middle-class values that modernization was supposed to bring to rural France. It looked as though within fifteen or twenty years, much of the farmland of Saint Cézert would be sold to outsiders like the Duparc brothers and Salut, or else it would be rented to what young farmers might still remain in adjoining communities. Either way, an era seemed to be drawing to a close.

4

Where Garlic Is King

As the crow flies, Cadours is about 10 miles west of Saint Cézert.[1] In fact, the two are naturally linked: le Margesteau, the brook that is nowadays pumped dry to fill up the communal irrigation pond of Saint Cézert, originates in the township of Cadours. And if the ambitious plans for a major lake in the Cadours area had not aborted, the waters of the Margesteau might very well be irrigating Cadours and its adjoining townships rather than Saint Cézert. Yet despite the proximity and this interdependence, Cadours belongs to a different world. Where Saint Cézert straddles the ancient flood plain of the Garonne river and the first terrace of the Gascon highlands, Cadours lies farther west in the heart of the Gascon hills. Not that these hills are very impressive: less than 300 feet separate the highest from the lowest point of the *commune* and there is hardly a slope too steep to be cropped. Where Saint Cézert has always had to contend with predominantly mediocre soils—gravel and *boulbène*—that are parched in the summer, the land in Cadours consists mainly of the same type of rich, water-retaining clay regionally referred to as *terrefort* that we encountered in Buzet's confined right bank area. Cadours does not include rich river-bottom, but it is relatively fertile soil, much less susceptible to drought than most of Saint Cézert's land. This good soil helps explain why Cadours has been and still remains a classic example of polyculture, of growing and raising a little bit of everything. This stands in contrast to the specialization of Saint Cézert, where for three generations vineyards and wine were a way of making the most of land that, until irrigation came along, was not much good for anything else.

The social structure of Cadours also differs sharply from nearby Saint Cézert, where, as we saw, a relative handful of substantial farm families had owned the bulk of the land at least as far back as the eighteenth century, if not earlier. Cadours had never known the same sort of polarization between landless laborers and employing farmers. In contrast to Saint Cézert, there were few day laborers or *maîtres-valets* in twentieth-century Cadours, just as there never were large, consolidated agricultural estates like La Mothe. It is true that well into this century, a number of Cadours farms were owned by middle-class or aristocratic landowners and worked by tenant farmers—sharecropping being the

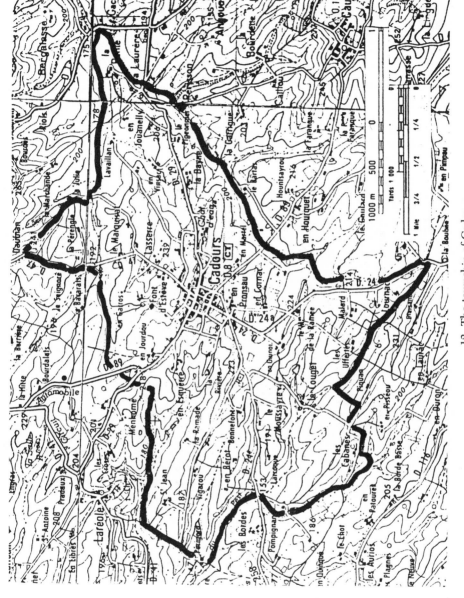

12. The township of Cadours

customary local arrangement—who, as late as 1936, comprised more than a quarter of all farm families. Yet they were outnumbered two and one-half to one by peasant smallholders who owned the land they farmed and most of whom got along without hired labor. Between the turn of the century and the period following World War I, there were just over ninety peasant families in Cadours, whose land holdings averaged about 12 hectares per family. Even today, with an acreage barely greater than Saint Cézert's and after decades of rural emigration and subsequent farm consolidation, Cadours still retains two and a half times as many family farms as its neighbor 10 miles to the east.[2]

Finally, the administrative fact that Cadours happens to be a *chef-lieu de canton*, a kind of minor county seat, sets it apart from villages like Saint Cézert, Juzet d'Izaut, and Buzet-sur-Tarn. Although the village, that is, the agglomeration of Cadours itself, has only about 500 inhabitants, it has not only its own post office and grade school, but a government tax collector, a branch office of the Crédit Agricole, grain storage bins for the regional producers' cooperative, a permanent post of the national *gendarmerie*, and, for the last twenty years, a *collège d'enseignement secondaire*, the French equivalent of an American junior high school. Moreover, thanks to an enterprising mayor of the 1960s, Cadours has had its own municipal water supply for the last two decades and even sports such a conspicuous symbol of urban development as an HLM (*habitation à loyer modéré*), a multistoried, subsidized housing project. The same former mayor and his schoolteacher wife, now both retired, have also taken the lead in organizing trips for local pensioners, who for years now have been gallivanting all over Europe by taking advantage of cut-rate group fares for senior citizens. Despite its small size, in some ways Cadours has the feel of a small town rather than that of a village.

Unlike the other communities that we examined, Cadours also differs in being a genuine market center. The weekly Wednesday market was first instituted in 1823, but it has only become regionally prominent during the last twenty years, paralleling the spectacular rise of Cadours' violet-colored garlic. Violet garlic has become to Cadours what sweet peas were to Buzet at the turn of the twentieth century, only more so. The weekly markets are particularly lively between July and December, when the garlic trade is in full swing. By attracting farmers from all over the canton, as well as from nearby communities in the *départements* of Gers and Tarn-et-Garonne, the Cadours market has in turn allowed a broad range of local shopkeepers and artisans to survive and, in some cases, to thrive. Because of its weekly market, the village is more than the mirror image of agriculture within the narrow confines of Cadours township boundaries. Cadours can count on its role as the focal point of a prosperous agricultural miniregion that has a specialty crop much in demand.

Cadours' amenities have in turn encouraged a fair number of Cadourciens who might otherwise have left altogether to commute to work in Toulouse. At the same time, a smaller number of people employed in Toulouse have chosen this village as their bedroom community, even though a daily drive of 35 kilometers each way is normally considered unacceptable. Although the phenomenon is not as common as in Juzet d'Izaut, there is even an occasional disused Cadours farmhouse that has been bought up as a summer and weekend home by Toulousains who prefer the accessible charm of the rolling Gascon countryside to the remote grandeur of the Pyrenees.

Although economically the village of Cadours seems to be holding its own and, in terms of population, even rebounding from its low point of the early 1950s, over the last fifty years outside observers have been generally unimpressed by the dynamism of its farmers. The picture presented is one of unrelieved backwardness and comparative stagnation, though evidently the historical context in which comparisons are made does keep changing. For example, when Dr. Suchich, who was to become mayor twenty-five years later, first arrived in 1932 to set up shop as the local veterinarian, he felt right at home. Having been brought up as the son of prosperous peasants in the Balkans, hardly an area in the vanguard of European agricultural progress, Suchich found Cadours at about the level with which he was familiar as a youth. Indeed, Cadours' peasants could not have appeared more immersed in their ancestral way had the Middle Ages been in full bloom. Much as the veterinarian welcomed the beginnings of agricultural change after World War II, even by the end of the 1950s when he entered politics, he did so in the hope of easing what he saw as his adopted community's traumatic transition to the modern world. Where other villages were progressing, he felt, Cadours still seemed to be floundering in its backward ways.

Another observer, Michel Barnadien, came to Cadours a quarter century later as a French expatriate from Tunisia, buying his farm in 1957. In central Tunisia he had been a large landowner used to American-style dry farming methods and motorized American heavy equipment. By purchasing 70 hectares of land divided between Cadours and an adjoining township, overnight he became the largest local proprietor. At the time that he arrived with his caterpillar tractor, he recalled—mistakenly as it turned out—there was only one other tractor in the village and that one owned by a local harvest contractor. Everyone else still relied on oxen or cows.[3] When Barnadien applied for a crop loan, a routine procedure in Tunisia, the local Crédit Agricole turned him down, because the members of the loan committee would not believe that he had really planted two-thirds of his land in wheat. Yet they steadfastly refused to see for themselves what he had done on the grounds that such an on the spot inspection was not the local custom. At Cadours, he was told, you put one-third of your acreage in wheat, one-third in fodder for your work animals, and one-third in fallow to rest the land. Barnadien was convinced that at the time of his arrival in Cadours, local agriculture was a full generation behind that of Tunisia.

The yardstick against which M. Marchette measured the local farmers was neither that of Balkan agriculture of the 1920s nor that of Tunisian dry farming in the 1940s. Marchette, the young agricultural extension agent for the canton who had only been stationed here since 1978, took many of the changes witnessed by Dr. Suchich and Michel Barnadien for granted, since everywhere, even in Cadours, French farming has come a long way in the last twenty-five years. Yet although he approved of the handful of local farmers who joined his study groups and showed interest in the garlic cooperative he was trying to promote, he found farmers in other villages of the canton far more receptive and progressive. Marchette did not enjoy preaching to the heathen, of whom large numbers were concentrated in the township of Cadours. At Cadours, he contended, the average farmer was likely to be a stick-in-the-mud, who muddled through in his own way, interested neither in making a special effort to shape up, nor in cooperating with others even if it was in pursuit of his own interests. At most, the local farmer was willing to accept ready-made solutions to his technical problems at the hands of the extension agent, but Marchette did not believe in handing out "prescriptions" like a pharmacist, even while admitting ruefully that doing just that would help fill up the waiting room during his scheduled office hours. Nevertheless, he insisted that his job was to help farmers take charge of their own lives, not to make them more dependent on outside authorities like himself.

Three generations of observers, yet the same expression, *une certaine mentalité* (a certain frame of mind) keeps cropping up in connection with Cadours farmers. Whether this state of mind ever really prevailed is hard to judge and impossible to verify. Had Dr. Suchich opened his practice in some other community of the French southwest, he might well have encountered conditions and attitudes every bit as "medieval" as those of Cadours. Barnadien, too, might have been equally shocked at agricultural backwardness had he bought his farm somewhere else in Haute-Garonne. As for Marchette, a young man still in his twenties, he tended to play down one aspect of local farming that seemed obvious to an outside observer: the fact that the average Cadours farmer was in his mid-fifties, notably older than the average age in more "progressive" villages of the canton, such as Garac, which have retained a greater proportion of the young. It stands to reason that, regardless of where they happen to reside, farmers who are only ten years away from retirement and who can count on no assured succession are unlikely to invest enthusiastically in agricultural modernization.

Even so, like farmers elsewhere in the French southwest, those at Cadours have been drawn into the age of mechanized equipment, larger farms, chemicals, high yields, and lagging prices. Perhaps it is true that, because they happened to be in an area where, as one local farmer noted complacently, "making a living is easy," they have not embraced modernity with passionate energy, as in Buzet-sur-Tarn, or been defeated by it, as in Juzet d'Izaut, or adapted to new opportunities with cool acumen, as in Saint Cézert. It may be that the course of agricultural modernization in Cadours has been marked by greater hesitation and reluctance than elsewhere. The favorite French catch-all adjective for forward-looking—*dynamique*—invariably employed to characterize the energetic go-getters of Buzet and commonly applied to the prudent ex-vintners of Saint Cézert, rarely comes up in connection with Cadours. Observers, and not least the Cadourciens themselves, are convinced that the locals are a bit lackadaisical. The same people who are quite willing to admit that most farm families that remain at Cadours have generally prospered over these decades of change are likely to attribute that success more to good luck than to deliberate effort.

Certainly there is little, if anything, in the historical record of Cadours and its agriculture to document this mythology of a unique local backwardness. In fact, there is so much in the story that parallels developments in Buzet-sur-Tarn and Saint Cézert (the mountainous setting of Juzet d'Izaut sets that community somewhat apart from the others examined in this book), that a detailed social history of Cadours would be redundant. It should suffice to concentrate therefore on a few variables that provide a basis for comparison.

One such essential variable is population growth and decline. The rough outline of the demographic evolution of Cadours will already be familiar from our account of other *communes*.[4] Cadours, like the other villages, grew throughout the first half of the nineteenth century, in this instance 31 percent, from 790 in 1800 to 1,033 in 1852. Thereafter it entered a slow decline, stabilizing at 800 to 850 inhabitants between 1891 and 1911. As in all three of the other townships, World War I proved a watershed: between 1911 and 1921 Cadours lost 17 percent of its inhabitants, among whom the twenty-three war dead constituted but a small fraction. In the next twenty-five years, it lost another 15 percent, stabilizing at just under 600 inhabitants between 1946 and 1954. Thereafter the township recovered rapidly: by 1977, with 776 inhabitants, Cadours had registered a population gain of 25 percent in less than twenty years and was nearly back to where it had been in 1800.

While these statistics may reveal some overall trends, they mask others that may have been every bit as significant. For example, the residence pattern within the *commune* changed markedly over the years: in 1809 of the 808 inhabitants, about 15 percent lived in the village proper, *le bourg*, another 15 percent in isolated farmsteads, and the remaining 70 percent in twenty-five hamlets. By the mid-1880s *le bourg* had 307 inhabitants, 35 percent of the *commune's* population.[5] Even though the overall population in 1921 was 13 percent less than what it had been back in 1809, the village itself had more than tripled, from 130 to 401 inhabitants: almost three-fifths of the township's population now resided in *le bourg*. Since 1954 this trend has continued: the recent growth of Cadours has been almost exclusively in and about the village. By contrast, the hamlets have disappeared altogether, leaving only place-names behind, the so-called *lieux-dits*, once shared by a number of households, but now usually referring to a single farm. Among a farm population that used to cluster, isolated homesteads have become the rule.

The population statistics also mask our ignorance as to how and why changes occurred. It would take a detailed study of Cadours' birth, death, and marriage statistics to get at the mechanics of population increase in the first half of the nineteenth century and its subsequent irregular decline and ultimate partial recovery. For example, the fact that at the height of the demographic boom in 1852 almost one-third of the inhabitants consisted of qualified voters, that is, of adult males twenty-one years old or over, suggests a society with too few children to even maintain its numbers. The earlier population expansion may well have reflected migration into Cadours rather than the drop in death rates that one might be tempted to surmise.

Drastic demographic decline in the first half of the twentieth century does not necessarily lend itself to traditional explanations of the "rural exodus" either: the rapid population decline in the decade between 1911 and 1921 that included World War I left the number of actual *peasant* households stable; there were just as many farms in 1921 as there had been in 1896, so that in the latter year members of agricultural families made up a higher percentage of the *commune's* population than they had twenty-five years earlier. Only from the 1930s onward is it possible to make a clear-cut, documented case of a net diminution of the local *farming* population.[6]

As at both Buzet and Saint Cézert, Cadours population statistics fail to take into account that some emigration was compensated by immigrants, who, as elsewhere, arrived in distinct waves that can be documented both by census lists and personal recollections. North Italian immigrants arrived as sharecroppers to take up vacant farms in the early 1920s and again in the mid-1930s. A second generation, either born in France or at least raised there, appeared after World War II, initially as tenant farmers. A fair number of Italian farm families settled permanently in Cadours, most of them coming to own at least some of the land they farmed. Some of these Italians have ended up among the leading farmers in the community. A second wave appearing in the early 1950s consisted of half a dozen northern families, mostly Breton, who had been officially encouraged to migrate from their own overcrowded *départements*. These Breton dairymen proved unable to master the constraints of local soil and climate, particularly when it came to producing fodder for their animals. All of them sold out within a few years. The third wave was made up of North African expatriates, who, beginning with Barnadien, started to buy land in Cadours in the late 1950s. They too had a difficult time adapting, although for different reasons than the northerners. Accustomed to vast acreages, extensive dry farming, heavy American equipment, and cheap, docile labor, in Tunisia and Algeria most of these *pieds-noirs* had been estate managers, rather than working farmers. In many cases, their commitment to a life of dirt-farming was tentative at best. Of at least five families who settled, only two were left in the *commune* by 1983 and only one, Barnadien, was deemed a success as a local farmer. Once the two remaining *pieds-noirs* retired, neither had offspring ready to take his place.

The patterns of immigration and the general population movement of Cadours are not unique. We have already met with similar, though not identical, patterns in Buzet-sur-Tarn. Both communities saw a relatively modest demographic growth in the first half of the nineteenth century, followed by a decline that stabilized the population at a lower, but still fairly high, level until World War I. Both saw considerable shrinkage until after World War II, followed by a rapid, and, for Buzet, complete, recovery. Both at Cadours and Buzet, waves of migrants took up vacant farms, with Italians playing almost identical roles in the two communities. However, unlike in Cadours, where the Breton dairy farmers were birds of passage, in Buzet the Aveyronnais dairymen became an integral part of the farming community. And thanks to the personal ascendancy of one youthful, dynamic *pied-noir*, Mallorca, among the young farmers of Buzet, the impact of the North African expatriates was much more marked there than in Cadours.

The similarity also extends to the two communities' recent demographic recovery. In Buzet's case, returning pensioners came to fill vacant houses in the village, while working couples from Toulouse built homes for themselves wherever they could get a good buy on a building lot. These two groups more than made up for the declining agricultural population, but did nothing to revitalize *le bourg*, which has become a kind of unofficial retirement colony. At Cadours, the turnaround also involved some Toulousains coming to buy or build houses for themselves, as well as the larger number of Cadourciens continuing to live in their home village, even while holding jobs in Toulouse. Yet unlike the situation at Buzet-sur-Tarn, beginning in the 1960s, demographic growth at Cadours also reflected its enhanced position as an administrative hub—the opening of the secondary school, for instance, brought a number of teachers with their families to settle in the community—and as a prospering market and district shopping center.

A second variable, the social structure of Cadours, and particularly the way land was owned and farmed, is more distinctive. More than either Buzet-sur-Tarn or Saint Cézert, Cadours has a long tradition of being a community where small peasant proprietors have owned much, though never all, of the agricultural land.

This does not mean that property was evenly distributed, any more than was local political power. For example, throughout the nineteenth century and down to the present day, Cadours' mayors have either been professionals or substantial proprietors living from landed revenues, rather than working farmers. No more than thirty different family names appear among nineteenth-century municipal councillors (almost certainly implying a somewhat larger number of distinct *families*) at a time when there were at least three hundred households in the *commune* and about one hundred adult males met the property qualifications for voting in municipal elections during the 1830s.[7]

Yet despite the prevalence of small-scale farming, Cadours never assumed the pattern of Juzet d'Izaut, with its desperately crowded and poverty-stricken peasants scrambling for an arduous living. As early as the middle of the eighteenth century, Cadours' peasants owned 42 percent of the land, while craftsmen, who were part-time peasants also in need of land to eke out a living, were in possession of another 9 percent. By comparison, noble landownership was not very significant, as four nobles owned only 14 percent of the land. Their position seems to have been somewhat weaker than at either Buzet or Saint Cézert. Bourgeois proprietors living from their landed income and such professionals as lawyers, notaries, and merchants owned a sizable share with 32 percent.[8] A generation later, by the time of the French Revolution, the basic pattern had remained unaltered, though the share owned by the nobility had edged up to 17 percent.[9]

From the nineteenth century on, the relations between social class and property holding are more difficult to trace, because modern land registers tend to be more reticent about social status than eighteenth-century tax rolls.[10] A few trends do emerge when the cadastres of 1839, 1913, and 1948 are laid side by side.[11] First, the proportion of land held in properties larger than 20 hectares remained remarkably stable throughout the entire period, hovering at around 36 percent of all land. Second, within that category, the largest estates tended to shrink over time: in 1839 the two largest had an acreage of 86 and 72 hectares respectively; in 1913, the most extensive estate was down to 59 hectares; by 1948, the two biggest properties comprised 49 and 38 hectares. Third, in 1839 all farms over 20 hectares had been owned by middle- or upper-class proprietors; by 1913, at least three or four were in the hands of working farmers; by 1948, about eight, that is, more than half of such properties, were personally worked by their owners. Fourth, the share of small landowners (with 4 to 20 hectares), most of them self-sufficient peasant farmers, tended slowly to increase. This group held 41 percent of the land in 1839, 43 percent in 1913, and 49 percent in 1948. Fifth, the share held by the most numerous group of proprietors, those owning properties too small to support a family (between 1/10 of a hectare and 3.9 hectares) declined imperceptibly from 23 percent to 20 percent of the land between 1839 and 1913, then more rapidly to 15 percent in 1948. Presumably, some Cadourciens sold off their modest patrimony after leaving their home *commune* for the city. To sum up: already a stronghold of a landowning peasantry in the eighteenth century, Cadours in the succeeding two hundred years witnessed the slow consolidation of peasant landed predominance.

We know less about a third important variable, changing agricultural practices at Cadours, than in the other *communes* in this study because very little reliable documentation has survived to shed light on agricultural conditions during the nineteenth century. What we do know suggests that Cadours's agriculture was perhaps somewhat slower in becoming market-oriented than that of either Saint Cézert or Buzet-sur-Tarn. This may simply have reflected the fact that nineteenth-century Cadourciens had even worse roads to contend with than the farmers of Saint Cézert. Buzet had, of course, always enjoyed a privileged location on the main highway between Toulouse and Albi and on the navigable Tarn River, as well as, after the mid-1860s, on two railroad lines. Yet whatever lag may have characterized its nineteenth-century development, during the twentieth century change at Cadours very much paralleled agricultural progress elsewhere in the region.

The main agricultural trends at Cadours will be familiar from our earlier accounts of other villages. From a tax account drawn up in 1776 we find out that land was allowed to rest every other year—probably also a common practice at both Saint Cézert and Buzet-sur-Tarn, though direct comparable documentation for those *communes* is lacking. We learn as well that the highest-rated of the three categories of land devoted to wheat growing in Cadours produced five times the seed sown, very close to the wheat yields recorded on the generally poorer soils of Buzet in the 1830s.[12] By the 1880s Cadours agriculture was described as including a wide range of crops: wheat, oats, barley, rye, corn, as well as beans, horse beans, peas, and lentils. Peasants relied entirely on natural meadows for fodder, at a time when ten miles away, farmers at Saint Cézert were already specializing in sowing fodder crops like clover, *luzerne*, and alfalfa. Even before the phylloxera struck, vineyards amounted to little, since suitable, light, well-drained soils were rare. Oxen and cows served as work animals; horses were used only for transportation. Wheat, poultry, cattle, sheep, and hogs were widely bought and sold on the weekly market, though Cadours, as already noted, lacked adequate road connections and was not linked to Toulouse by narrow-gauge railway until 1903.[13]

The major agricultural changes of the early twentieth century seem to have centered on even greater diversification, with the large-scale introduction of fodder crops like clover, *luzerne*, alfalfa, and vetch and of tubers like potatoes and Jerusalem artichokes. These, in turn, permitted a more sophisticated crop rotation that must have cut down on the acreage left fallow at any one time.[14] Yet crop yields were still closer to the norms of the eighteenth century than to what were to be those of the late twentieth: the "fairly good" year of 1921 produced 1,440 kilograms of wheat per hectare (roughly seven times what had been sown); a "poor" (*médiocre*) year (1913) had yielded 960 kilograms, just under the 1 to 5 crop ratio considered average for good land in the eighteenth century. By 1913, the annual agricultural statistics also listed 10 hectares planted in garlic, the first mention of a specialty crop that, some fifty years later, was to make Cadours a name to reckon with on French produce markets.[15]

Changes in agricultural tooling and equipment seem to have begun around the turn of the twentieth century. For example, one farm of 12 hectares run by sharecroppers acquired its first reaper in 1902. At that time wheat was still threshed by oxen pulling a circular roller, soon to be replaced by a threshing machine powered by four yoke of cattle. Around 1908 the stationary steam engine displaced the animal-powered threshing machine. Between the wars a mobile steam engine, a sort of locomotive needing no rails, in its turn replaced the stationary engine, just as reaper-binders displaced the earlier simple reapers when it came to harvesting. All this represented a familiar evolution that we have already encountered elsewhere. Cadours may have begun to evolve a few years

later than Buzet and Saint Cézert, but the general pattern of technological progress was very much the same.

The one original feature at Cadours was the role of a farmers' union,
founded in 1902 on a cantonal basis by some one hundred farmers. Initially promoted by an outsider, a professor at what was then the agricultural institute at Purpan, the union emphasized technical progress and
self-help. Since Purpan was a Catholic institution and 1902 was a time of
acute conflict between church and state, the new organization may have
had indirect political aspirations. If so, they have left no record. The
farmers' union's major early achievement, it seems, was to introduce and
to popularize Brabant-type plows, which meant that by the interwar period, the old, light, wheelless plows already familiar to the Romans and
its somewhat more advanced wheeled version had been relegated to minor chores: two yokes of oxen pulling a Brabant had become the new
norm. The farmers also founded a credit union and organized such activities as visits to model farms in adjoining *départements*. After World War
I, the cantonal union bought shares in the newly organized Farmers' Cooperative of the Southwest and moved from promoting Brabant plows to
popularizing disk harrows.[16]

The agricultural survey of 1929 suggests that the union had some impact. Although there were, as yet, only three disk harrows in Cadours,
the survey listed sixty Brabant double plows, fifty-six horse-rakes, and
even five tractors with as many plows with multiple shares, although
prior to World War II such motorized equipment was generally owned
by agricultural contractors rather than by ordinary farmers.[17] If anything,
Cadours had pulled slightly ahead of Saint Cézert and Buzet-sur-Tarn
when it came to the use of modern farm equipment. By the 1940s and
1950s the farmers' union devoted most of its energies to organizing bulk
purchases of chemical fertilizers and seed as a way of saving its members
money. Since that time the organization has declined.[18]

Everywhere in the French southwest the great modernizing spurt—at
Cadours the decisive years seem to have been between about 1955 and
1965, although the precise limits of that modernization did not shape up
until the early 1970s—may be thought of as taking place within a sort of
modernizing nebula, a cluster of constraints and possibilities that impinged on each other to the point where cause and effect are hard to
distinguish. Many of these constraints and possibilities were national and
even international in scope and to this extent all of the townships studied
in this book were subjected to the same forces.

We have already seen from our examination of Buzet-sur-Tarn, Juzet
d'Izaut, and Saint Cézert that both limitations and opportunities were
variously perceived, because responses varied with local geographical,

social, human, and historical differences. In discussing the process at Cadours, it would be pointless to rehash the many aspects of agricultural modernization already sketched in connection with other villages, aspects in which Cadours evidently shared. There was, for example, much in common among rural townships of the French southwest when it came to the link between the introduction of the tractor, the expansion of farm acreage, and, initially, the expansion of stock raising and dairying. Yet rather than elaborate on what has already become familiar, I would like to focus on responses that were peculiar to Cadours, as well as on wider issues emerging more clearly in this township than in other communities studied.

In trying to make sense of agricultural modernization at Buzet-sur-Tarn, I sought to demonstrate how decisive a role the tractor had played in this modernizing constellation. In fact, its advantages on Cadours farmland may have been even more self-evident than on the thin soils of Buzet and Saint Cézert or within the increasingly pastoral economy of Juzet. Because land at Cadours was heavy loam with a thicker layer of topsoil and because a tractor could plow two to three times as deep as a plow pulled by cattle, for a few years motorized plowing was a means of tapping the bounty of virgin humus that the old plows had not reached. Even the least progressive peasant at Cadours could seize upon this advantage. Of all farmers active in Cadours, Barthez was the very last to acquire a tractor. Yet for several years before he finally did, he helped his neighbor with his chores in exchange for having his fields plowed by the latter's tractor, because deep plowing was such a painless way of increasing the harvest. Here, as elsewhere, the tractor also served as the Trojan horse of agricultural modernization.

Yet not even at Cadours was the tractor capable of initiating this revolution all by itself. This is neatly demonstrated by what might be called the "premature motorization" of the Saint Antonin farm. The late Saint Antonin—deceased parents are invariably referred to as "my poor father" or "my poor mother"—the father of the farm's present head, bought a tractor as early as 1937, at a time when only harvest contractors owned such equipment, which they used chiefly for powering threshing machines. Saint Antonin's decision was motivated by what was universally seen as terrible luck, namely, the outbreak of hoof-and-mouth disease, the dreaded bovine plague, in his barn. He had no choice but to slaughter all his work animals and, apparently, the quarantine rules were so strict that he was obliged to wait several years before replacing them. His only solution was to buy a tractor for plowing, a clumsy, secondhand machine with metal lugs instead of tires. Unfortunately, no equipment specially designed for use with the tractor was available: regular plows had to be haphazardly adapted. The venture was not a great success and the machine itself a nuisance that tore up the roads. After three or four years, when the quarantine on his barn was lifted, Saint Antonin got rid of the unwanted tractor and went back to plowing with oxen. Evidently there was nothing magical in owning a tractor per se when it was introduced in unfavorable circumstances: the tooling that would have shown off its advantages being unavailable, this involuntary experiment in motorized agriculture made no positive impact on Cadours farming whatsoever.

The cooperative ownership of sophisticated new agricultural equipment, on the other hand, played a central role at Cadours that it did not play in our other communities. Parallel to the network of agricultural extension agents put in place by the mid-1950s through the joint efforts of the Chambre d'Agriculture and the General Council of Haute-Garonne, a separate team of technicians was sponsoring municipal CUMAs (Coopérative d'utilisation de matériel agricole, that is, a coop for the use of farm equipment). The regional socialist leadership, strongly represented in both organizations, favored such a municipal approach to modernization, offering financial incentives for such cooperative groupings. The CUMA at Cadours was one of the earliest such ventures in the *département*.

Originally the creation of the Cadours CUMA was prompted by the conversion of local farmers to the use of combines, which consolidated harvesting, threshing, and bagging grain into a single operation. Because the new combines were extremely expensive and harvest contractors understandably cautious, when combines were first introduced in the mid-1950s they were in short supply. "They promised they'd be here in three days and it would take them three weeks," was the general lament in Cadours. In the old days, with sheaves carefully stacked to repel rain, such delays were no more than a nuisance. Since the combine, on the contrary, harvested and threshed in the fields, half the harvest could be lost if wheat on the stalk was exposed to early rains for several additional weeks.

This was the practical problem that led Luigi Pascaglia and two or three other progressive young farmers to organize the local CUMA. At the same time, they seem to have been socially conscious as well, hoping that the cooperative would foster the kind of solidarity formerly created by the *colle*, the team of neighbors who worked together on the mobile threshing machine. For this reason they insisted that membership in the new CUMA be open to all who wanted to join. In fact, when this equipment coop went into operation in 1963–1964, well over half of the township's farmers had adhered to one or more of its sections. Each section was organized around a particular piece of agricultural equipment, for, besides the combine, obviously the pièce de résistance, the CUMA had also bought a manure spreader, one seed-drill for wheat and another one for corn, a fertilizer spreader, and the like. At its peak, the organization had well over thirty members and its combine section alone began with as many as seventeen. And even though it hired two full-time drivers/ maintenance men for the seasons during which the equipment would be employed, the use of the equipment turned out to be notably cheaper than the rates charged by contractors.

Yet despite the financial advantages that it could offer its members, the Cadours CUMA foundered within ten years, only to be reborn on a much more modest scale. Some of its problems were the result of inexperience, others reflected a massive ignorance as to the nature of the financial commitment that such a cooperative entailed. Finally, and these are the problems that former members still like to evoke with sheepish amusement, there were the obstacles posed by the famous local mind-set (*la mentalité des gens*), the excessive individualism and lack of social discipline that has been the bane of many a French reformer. If the example of Cadours amounted to more than a local abberation, there may be good reasons why appeals to social solidarity are voiced with greater fervor and frequency in France than in the rest of the Western world.

Yet perhaps more should be attributed to some initial practical mistakes in the way that the new equipment cooperative was launched. The first problem was caused by the fact that the CUMA began its career by buying an underpowered combine that was too small and too cheap; it was therefore too slow-paced and broke down far too often. The choice of this machine had been dictated by older members' reluctance to take on too great a debt. They also pointed to the miniature size of so many of Cadours' fields that would not have admitted a larger piece of equipment cutting a wider swath. Even the small combine had trouble turning around or negotiating the trees scattered on some of the fields belonging to various members.

Nor could the organizers know that some pieces of equipment simply do not lend themselves to collective ownership. There are certain agricultural tasks that must be undertaken without delay, while collective ownership implied waiting one's turn. Wheat or corn must be sown when the fickle weather is just right and the soil is not too mushy. Nitrogen must be spread or sprayed at a very precise stage of a plant's growth and not ten days earlier or a week later. In such cases, there was no equitable way of passing scarce equipment around to the membership.

Finally, experience in the French southwest since 1963 has shown over and over again that a CUMA gets along only if its members, as individuals, get along with each other. However appealing unlimited democracy may be in theory, the reality of village life is such that family feuds are *not* laid to rest when lifelong enemies become joint owners of a vital piece of machinery. Such feuds are then merely carried on by other

means. In practice this has meant that in the long run this sort of cooperation is usually only workable within small, compatible groups. An
equipment cooperative is likely to work among a bunch of partners at the
weekly card game, among former schoolmates and friends, in short,
among people who are already compatible before they organize. These
were regional facts of life to be learned. As pioneers, the cooperators of
Cadours were bound to learn them the hard way.

Until several years later when Allier, the extension agent, tried to explain it to them, most members of the CUMA had only the dimmest idea
as to what cooperative ownership really involved. The notion that co-own-
ership was proportional to use, and therefore to acreage, made some
dent. Much more difficult to grasp was the idea that, while utilizing a
piece of jointly owned machinery, members not only had to defray current maintenance costs and the repayment of the loan contracted, but
they also had to worry about maintaining adequate working capital and
building up the amortization fund that would ultimately pay to replace
worn-out equipment. This lack of understanding was translated into a
widespread reluctance to pay the assessments voted by the membership.
As Pascaglia recalled, one-third of the members always paid on time, another third paid "after you pulled them by the ear," and a last third could
never afford to pay at all "as they drove off in the brand-new [Citroën]
Deux-Chevaux they had just bought for themselves."

This irresponsibility was magnified when it came to deciding the time-
table and order in which a particular piece of equipment would be used.
Farmers who groused but accepted the delays imposed upon them by a
commercial harvesting contractor—who, like God and the weather, was
beyond their control—were unwilling to brook the slightest delay now
that they were co-owners of a combine. There was always an intense reluctance to go first, since the grain might not be quite ripe, but once
harvesting had begun, almost everyone insisted that they had to be next,
that their wheat was riper than their neighbor's, that they were being
picked on, discriminated against, treated with disrespect. What also had
to be taken into account was the changeable weather and the predictable,
yet unforseen, delays caused by machine breakdowns. In what was not
Cadours' finest hour, the squeakiest wheels tended to get greased first.
It was *la pagaille*—a donnybrook.

In retrospect, Cadours' adventure in mass cooperation still evokes wry
grins among its former participants, yet whatever its shortcomings, financially the CUMA did ease its members' transition to modern farming.
The experience itself was both exasperating and disillusioning, yet by fur-
nishing essential services at very low cost, the CUMA provided what one
former participant called "a bubble of oxygen." Small farmers in particular were given access to expensive equipment indispensable for the new

agriculture without the strain of having to take on too great a burden of debt too rapidly.

Former members were amused and incredulous to learn—for most, the Cadours CUMA had become ancient history—that twenty years after it was purchased, the little old combine, which they remembered as held together by baling wire, was still running. For the four remaining CUMA members, all of them aging farmers with very small acreages, the old and by now long amortized combine reduced their harvesting costs by a factor of four or five. Not only did this saving make a big difference in their modest farm operations, but the four survivors had learned to work together without much friction. In a backhanded way, their modest success had validated the faith in the power of human solidarity that animated the socialist politicians and the pragmatic young local idealists twenty-five years earlier.

Personal recollections at Cadours also illuminate the relationship between agricultural modernization and changes in life-style and family relationships that a priori would seem only remotely linked. Despite the gradual agricultural improvements since the beginning of the twentieth century that we noted, on the eve of the great leap into modernity, the traditional round of peasant life seemed hardly affected at all. Outsiders like Barnadien were struck by earthen floors, fireplaces used for the family's cooking, the absence of running water and indoor plumbing. Electricity, by the 1950s, had long been available in most of the Cadours countryside: its introduction dated from the period of 1925–1932 and only a few outlying farms had to wait till 1960 to be connected. Laundry had to be done by hand as a matter of course. Cadours farmers were perfectly familiar with cars from at least the early 1930s on, but the half dozen motorized vehicles locally owned at that time belonged to middle-class professionals like the doctor and the veterinarian, although during the war at least one farmer was reputed to be hiding a car of his own that might otherwise have been requisitioned by the Germans. For the first several years after 1945, even the *motos*, really motorized bicycles, remained something of a novelty in peasant households. Certainly as late as 1950, for Wednesday's weekly market, people arrived by horse-drawn wagon, by buggy, by ox-cart, by bicycle, or else on foot, with or without a pushcart or wheelbarrow.

M. Allier, the first agricultural extension agent in the canton of Cadours, was well aware that introducing some daily comfort into the lives of farm families (refurbished or rebuilt homes; kerosene, gas, or electric stoves; refrigerators; washing machines; showers; indoor toilets; mobylette first, family car later) could not be separated from tractors, hybrid corn, new strains of wheat, seed-drills, chemical fertilizers, herbicides, and all the rest. In this respect, *pieds-noirs* like Barnadien, who considered domestic comfort his due and was willing to spend good money to bring his house up-to-date, became highly visible and effective propagandists for modern living conditions. Only by replicating the paraphernalia of contemporary urban comfort throughout the countryside might the young, and particularly young women, be induced to stay. And without a solidly implanted younger generation, agricultural progress was doomed.

Even though by the late 1950s the Crédit Agricole loosened its purse strings to finance home improvement loans of various sorts, the extension agent saw one of his main tasks as finding ways for local peasants to put money aside for such improvements. For a while, as in Saint Cézert, tobacco was the cash crop encouraged. As the demand for American hybrid corn exploded by the middle of the 1950s he induced some of the younger farmers of Cadours and nearby *communes* to organize the cultivation of hybrid seed corn, which was more profitable and less labor-intensive than tobacco. However, the effort had to be collective, because corn grown for seed must be kept isolated from other corn fields by a distance of at least half a kilometer.

This local specialty crop (nowadays still grown in nearby Garac, but no longer in Cadours itself) allowed those farms where the older generation hoped to retain their children on the land to enter upon a period of feverish domestic improvements. By the early 1970s most of the standard urban household amenities, from central heating to television, were widely diffused among Cadours farmsteads. It was only the poorest farmer who still lacked a car and even he was sure to have his mobylette. The most obvious gap in living conditions between city and country had been bridged. If some of the young continued to leave for Toulouse, they did so because they felt cramped or bored at Cadours, or because they preferred putting in their thirty-nine hours on the assembly floor of the Aerospatiale or selling notions at the Nouvelles Galeries, to longer hours atop a tractor or force-feeding recalcitrant ducks. The young no longer left, as many had earlier, merely to escape medieval drudgery.

Shrewd agricultural technicians, like Allier, also soon became aware that agricultural modernization entailed a "family process," a radical restructuring of familial power relationships. As long as farmers remained outside of the national social security system and had no old-age pension to which to look forward, the best strategy of old people aiming for security was to cling to patriarchal (or, in the case of a widow, matriarchal) authority by hanging on to the title of their farm.

Traditionally, one or more young adults worked as full-time, unpaid help on the parental farm, as *aides familiaux* (family helpers) as they were officially known. In the 1940s Jean Merle, already a young man in his late twenties, had to look for the occasional odd job like splitting wood to earn the pocket money that might allow him to take out his girlfriend. Working full-time on his parents' farm, he never saw the color of money. He was no anomaly at Cadours.

Generational conflict had always been built into this system, but accelerated modernization in the 1950s made the conflict more abrasive. In some cases, as in the Renal family, three generations lived under one roof, with the real authority resting with an autocratic grandfather rigidly opposing all agricultural innovation. In this instance, the middle generation, that is, the father, sided with the grandfather's conservatism, while the grandson rebelled against tradition. In like circumstances, most youthful rebels would take off for Toulouse. In this instance, the defiant young man married and set himself up as a tenant farmer in a neighboring township, only returning when he finally inherited the family farm. Yet it took the death of the old patriarch and the retirement of the father before the family situation could be normalized and the younger generation was allowed to admit the twentieth century to the family homestead. In turn, the retirement of the father was facilitated not only by the officially mandated retirement system finally instituted in 1955, but by the IVD (*Indemnité viagère de départ*) legislated in 1962, providing an attractive supplement to the regular pension to those willing to turn over their reins to a young farmer as soon as they reached sixty-five years of age.[19] Both father and son were beneficiaries.

As most members of the older generations were less mulish than father and grandfather Renal, the changing of the guard was usually less traumatic. We already saw that one of the factors that permitted the successful agricultural transformation of Buzet-sur-Tarn had been the fathers' willingness to initiate changes demanded by their sons—mechanization, farm enlargement, land consolidation—to encourage the younger generation to take up their succession. The same softening of the patriarchal model occurred at Cadours, if only by a variant of natural selection: farms headed by old-style authoritarians were likely to be doomed because resentful sons would simply up and leave. Those farmers (and their family farms) survived who learned to cooperate with their sons, which usually meant some degree of informal power sharing that would have been inconceivable a generation earlier. Given the beckoning urban alternative, young people could no longer be coerced.

This also altered the role assigned to children and adolescents on the family farm. The farmer who wanted his son to carry on—and not all of them did, as we saw at Juzet d'Izaut—felt that the days of exploiting familial child labor were gone forever. For the first time in the history of agriculture, schoolwork was seen as the sole responsibility of school-age children. At Cadours and probably elsewhere this attitude seems to have been carried to what American family farmers might consider absurd lengths: local farmers made a point of contrasting the farm work that they themselves had been asked to shoulder while still at school, with their children's complete exemption from any productive labor—usually to age sixteen, but often to eighteen. This deliberate insulation of farm children from the daily realities of running a farm was probably self-defeating, if it was meant to encourage the young to remain on the farm. The first generation to routinely attend agricultural secondary schools was therefore, paradoxically, also the first to lack any firsthand experience of farming that would have made its schooling meaningful. Yet in this process of catering to its potentially footloose offspring, within half a lifetime, the traditional, patriarchal farm family had moved a long way in the direction of an overprotective and child-centered French middle-class model.[20]

This transformation of the farm family linked to the massive changes in agriculture also had other significant aspects. Even prior to World War II, the rural family had already been visibly undercut by a shortage of marriageable women. This resulted because a larger number of young women than of young men had been leaving villages like Cadours for city living, unwilling to follow their mothers into what they saw as the harsh and unrewarding life of farm wives. Even before modernization took off in Cadours, the increase in the number of bachelor farmers—*vieux garçons*, old boys, as the French quaintly call them—was short-circuiting normal, hereditary transmission of landed property. During the 1950s young women still remained in the van of the rural exodus, so that another generation of remaining young farmers had a hard time finding wives. Jean-Marie Abadie, for instance, the most successful young farmer at Cadours (the fact that at age forty he was still locally classified as "young" is a commentary on the graying of the farming community, a related problem), was married and had children, but the majority of his friends from the regional agricultural high school he had attended remained single. It was taken for granted that no young woman in her right mind would choose to marry into a farm that was backward, impoverished, or too small. Keeping up with progress thereby became not simply an economic or cultural imperative, but a prerequisite for finding a mate and enjoying a normal family life. At the same time, success as a modern farmer only bettered the odds of finding a wife: it provided no guarantees.

Therefore young women inclined to become farm wives, or even to marry farmers without giving up their own jobs or careers, have come to have immense bargaining power. Colette Abadie is a case in point, though as a couple the Abadies are too formidably capable to be depicted as typical. A farmer's daughter from 10 miles away, she came to Cadours after graduation to work as secretary to a local contractor. When she and her fiancé, Jean-Marie, set the wedding date, she insisted that she had no intention of letting herself in for *la cohabitation*, sharing one household with her in-laws. In their case, this stipulation posed no problems, since Jean-Marie, aided by parents who were both prosperous and unusually understanding, had acquired a farm of his own at the age of twenty-four, the only such case in Cadours. After a first child arrived, the mother-in-law was helpful enough to babysit her grandchild, while Colette stayed at her secretarial job. When a second child was born, the young woman quit, convinced that she could earn more on the farm without spending all day away from her family. In retrospect she had no regrets over her decision, though hers was the only case of an employed wife of a farmer giving up her outside job that I encountered.

The Abadies farm 60 hectares, the largest farm located entirely within the township boundaries of Cadours. Jean-Marie is responsible for all of the crops—wheat, rapeseed (*colza*), sunflowers, but also a couple of hectares of garlic. The sale of major crops goes mostly to defray day-by-day farm expenses, taxes, debt and interest payments to the Crédit Agricole, and depreciation and replacement of equipment. As soon as he brings the garlic into the shed, it becomes Colette's sole responsibility. Between August and October, she is in charge of preparing the garlic for the market, peeling 160 kilograms a day herself and supervising the vacationing adolescents whom they hire to pull and peel garlic. By October 15, she is done. On November 1, she begins to force-feed her eight hundred ducks and geese, the work of late fall and winter that consumes six hours of her day. Because it is such an onerous job, no one else in Cadours stuffs more than three hundred fowl. Between Colette's 8 tons of garlic (in 1982 the drought cut the quantity, but raised the selling price) and her eight hundred ducks and geese with their precious enlarged livers, she must have grossed some Fr 300,000 that year. After covering her considerable expenses, she will have been left with what amounts to the couple's net income, handsome even by French upper–middle class standards. Both husband and wife are very hard workers nakedly proud of their success. This is why they resent being razzed as "rich peasants" whenever they join the secondary school teachers and other young professionals on their weekend ski trips to the Pyrenees, the only local farmers to go on such outings. And when I interviewed them, they spoke with equal confidence and forthrightness, a sociology textbook example of marital equality. The Abadies may be exceptional, but they probably also are straws in the wind, a hint of new egalitarian farm families in the making in what used to be the unchallenged domain of traditional Mediterranean patriarchy.

In retrospect, as I already noted, the decade 1955–1965 appears as the period of Cadours' agricultural takeoff, yet the profound changes in the

making were much less obvious to contemporaries. Population decline,
French conventional wisdom had long proclaimed, was tantamount to
decadence and this was the time when Cadours' population was still falling to what turned out to be its low point. If some farmers seemed in the
process of modernizing, many peasants of the older generation who did
not have sons at home to push them, remained mired in archaic methods.
Local society seemed frozen in a premodern mold. That the diehards
would gradually die off and their land pass on by purchase or lease to
modern-minded farmers seems obvious only from a much later perspective. The municipality of Cadours itself was anything but progressive,
with the incumbent mayor throwing up his hands whenever anyone
broached the idea of changing things.

   This is the background against which the mayoralty of Dr. Serge Suchich must be understood. By 1959, when he became mayor by defeating
the slate of his elderly predecessor, Suchich was a man in his early fifties
who had spent the last twenty-five years in Cadours. As the area's only
veterinarian, he knew the region and its people. There is no reason to
doubt his genuine affection for the local peasants and his devotion to his
adoptive hometown. A lifelong socialist, Suchich was also a paternalist by
personal inclination, though he would have been shocked to be so described. A man of great energy, integrity, pride, and self-confidence, he
tended to become irascible when opposed. Nor was he much plagued by
doubts as to what was good for his *administrés*, an untranslatable term
only feebly rendered as constituents in American English, but one that
in French also carries certain hierarchical overtones.

   He saw himself—and rightly so—as a man who got things done.
Within the centralized French system in which Suchich had learned to
operate, there were traditionally and politically accepted ways of going
through channels. A rural mayor who had problems at the departmental
level, whether with the General Council, the prefect, or with some
agency under their respective jurisdiction, consulted the *Conseiller général*, the elected cantonal representative to the *département*. In the
case of the mayor of Cadours, this would have meant going through the
dean of the General Council, Jean Delile, who had represented the canton of Cadours ever since 1912.[21] If his *commune's* problems could only
be resolved by Paris, a rural mayor would be expected to look up the
local deputy to the National Assembly, a gentleman who also doubled as
mayor of Colomiers, one of the mushroom suburbs of Toulouse. The
problem was that both the general councillor and the deputy were so-
cialists, while the national government was Gaullist and conservative,
with the local prefect responsive to that same politically unsympathetic
regime.

   Going through channels under these circumstances was therefore not
very effective. Whenever Suchich felt stymied, which happened often,

as one of his former *administrés* summed up admiringly twenty years after the event, the mayor "with bag and baggage" (*avec ses cliques et ses claques*) took the train to Paris to negotiate with the real power brokers, the permanent civil servants staffing the upper reaches of the bureaucracy, men who prided themselves upon being above politics. And Suchich discovered that they responded to reason, enthusiasm, and sheer persistence.

This unorthodox approach may not have endeared him to the incumbent socialist politicians he bypassed, but to most of his constituents he was the mayor who brought home the bacon. "Getting things done," in the context of French local government was at least as much a matter of tapping official subsidies—rural municipalities could play that game as much as farmers did—as it was obtaining the authorization to do what needed to be done. During the first years of his administration, Suchich was able to deliver on a municipal water system extended to every isolated farm, the completion of rural electrification, the building of a new grade school and of the first regional Collège d'enseignement secondaire (commonly known as CES), as well as the construction of a subsidized, low-rent housing project.[22] We have already mentioned some of these achievements earlier in talking about the features that gave Cadours a certain urban tone. Suchich was particularly proud of the municipal water works, because aside from meeting a desperate need, without increasing taxes, he had thereby managed to endow the municipality of Cadours with a steady income from water rates, granting it the ability to finance further improvements. These achievements pretty well exhausted the platform on which his slate had originally run for control of the municipality.

Land consolidation for Cadours had not been part of that platform and originally land consolidation was not on Dr. Suchich's mind at all. Having fulfilled all of his electoral promises, he was looking for new worlds to conquer in behalf of the population of Cadours. Taking vacation trips to his Balkan homeland, he had driven through the hill country of northern Italy, a landscape that sometimes reminded him of the Gascon hills. What had struck him in Italy were the man-made irrigation ponds dug to catch the runoff from the hillsides. These hillside ponds (*lacs collinaires*) provided ingenious, reliable, and low-cost irrigation reservoirs. If the Italians could irrigate in this way, why not the Cadourciens, whose agriculture faced drought every four or five summers? Moreover, it was intolerable that a country like France should so lag behind. He gradually became determined "to give his peasants irrigation."

It was characteristic of the mayor that immediately upon his return, he resolutely began to look around for a suitable hillside site in Cadours. It was equally characteristic that he did not apprise his municipal council-

lors of his plans or that he made no attempt to find out whether his constituents were even remotely interested in irrigation. He saw himself as
a would-be benefactor striding into the sunrise of the modern age, not as
a local politician looking to his rear for support and approval.

By 1963, his idea no longer seemed quite so visionary, for the first
irrigation projects in the French southwest were being underwritten and
in Cadours itself, one enterprising young farmer, André Reulet, had
dammed up a spring on his land to create a hillside pond with which he
began to irrigate his fields with a network of sprinklers. The mayor of
Cadours tried to keep abreast of such developments by corresponding
with the officials of the Ministry of Agriculture in Paris. One fine day,
having just attended a mayors' conference on irrigation in neighboring
Tarn-et-Garonne, Suchich had occasion to corner a high official of the
Génie rural, the bureaucracy concerned with agrarian improvements.
The bureaucrat waxed enthusiastic about the potential of irrigation, but
when the mayor of Cadours produced a detailed map of his *commune*, he
was told point-blank: "Irrigation will never work with fields that tiny and
dispersed. Before you can even think of irrigation, you must have your
township's lands consolidated." This was the gist of the dinner table conversation in 1963 that launched *remembrement* in Cadours.

The notion that dispersed plots ought to be consolidated into larger fields
suitable for modern equipment had surfaced in Cadours at least as early
as the mid-1950s. Although the first generation of tractors was small
enough to adapt to the traditional patchwork of small fields, as we noted
earlier, even the first modest combines often had trouble turning around
in these same fields. Pascaglia, having arrived at Cadours as a tenant
farmer in the 1950s, took stock of the problem as soon as he had succeeded in purchasing the farm he had previously leased. He began his
career as a proprietor by buying out an enclaved meadow belonging to a
neighbor before proceeding to other steps to make his work more convenient. The one who attracted the most attention in the village was Barnadien, the *rapatrié* expelled from Tunisia, who, by spending what to
the locals seemed a fortune, bulldozed hedges and bought up land to end
up with well over 50 hectares all in a single block of land. There had
previously been twenty separate parcels marked on the Cadours land
register.

Pascaglia and Barnadien, as well other North African expatriates, may
have been more single-minded in adapting their field system to modern
needs, but a number of long established local farmers also made substantial progress, mostly by exchanging intermingled fields with neighbors,
"friendly exchanges" that, as we noted, were exempted from the otherwise mandatory real estate transfer tax. By the time official land consoli-

dation was launched, a number of local farmers had reduced their scattered fields to a single "island" of land, sometimes with one or two additional outlying fields too far from their farm to incorporate. In fact, by 1966 the number of plots officially recorded on the land register had declined by one-third over what it had been a generation earlier. This improvement, however, still left 1,887 legally distinct fields parcelled out among sixty-six farmers (but 238 landowners). Even after all these efforts, the typical Cadours farm still had twenty-nine separate fields, each of which averaged just over half a hectare, little more than one acre.[23] Yet if farmers were far from apathetic about this obstacle, neither was there a ground swell for drastic change.

The process of officially-sponsored land consolidation at Cadours turned out to be a near-disaster, though even at its most mismanaged, such a procedure has some positive practical results. At the same time, the drama should not be exaggerated. Lawlessness, despite tough talk about "letting the hunting rifles speak," amounted to no more than a bunch of irate farmers pulling out new boundary markers placed throughout the *commune* by the surveyors. The resistance to *remembrement*, although it was embodied (as at Buzet, but with broader support) in that most typically French phenomenon, a formal association organized to keep something from happening, was only able to slow the process, helping to drag it out over eight years.

Still, Cadours did provide adequate source material for a "how not to" manual on land consolidation. Because the mistakes made and controversies engendered carried over to other issues that in turn changed the direction of the *commune*'s agricultural evolution, some analysis of just what went wrong and why seems in order.

We saw that in the mayor's mind land consolidation was not an end in itself, but a means to making large-scale, communally organized (remember this was also the time of the communal CUMA) irrigation practical. Even from Dr. Suchich's own "official" account, it is not at all clear that he kept his municipal councillors apprised of his plans and ideas. As he recalled it, the possibility of land consolidation was first brought up on its own merits at a municipal council meeting in January 1964. Since, in retrospect, the mayor admitted that he preferred having the issue raised by someone else, presumably the two councillors who did bring up the subject had been taken into the mayor's confidence.[24] Moreover, Dequilhempey, the councillor who initially put forward the proposal to request *remembrement* had a very personal interest in having the municipal roads redrawn, a normal corollary of land consolidation. He and his neighbor Perrot had recently been involved in a bitter squabble over a right-of-way that Dequilhempey claimed and that Perrot, a North African *rapatrié*, had challenged. Somewhat imprudently, the mayor and the municipality had taken sides with Dequilhempey, whose legal claim was shaky. Providing access for Dequilhempey by laying out a suitable municipal road was a painless way of solving what had become a divisive question. Dispaut, the second proponent, was a harvest contractor by trade. Having, no doubt, much firsthand experience with the incompatibility of Cadours' crazy-quilt field system with the new combines, he too had a much more vital and personal stake in drastic change than the average local dirt farmer. Apparently the link between land consolidation and eventual irrigation was not discussed at the time, though in succeeding years Dr. Suchich made no secret of his plans.[25]

The strategic mistake—or, according to opponents, the original sin—of land consolidation at Cadours was that the mayor and the municipal council officially went on record in favor of *remembrement* before consulting the community's farmers and landowners. An informational meeting to which farmers were invited was indeed held in April 1964, but a month earlier the municipal council had unanimously voted not only to make a demand for land consolidation to the authorities, but, upon the mayor's recommendation, they had also selected the more thorough (and authoritarian) *remembrement* over its milder alternative, *la réorganisation foncière*, a reallocation of fields amounting to an officially supervised multilateral exchange on a voluntary basis.[26]

Weeks before the April meeting, word had gone out that "Suchich is going to do a *remembrement* at Cadours." Not surprisingly, a good many farmers who attended that meeting felt that they were confronting a fait accompli, that they were being informed, not consulted. The distinction was crucial. Although it was legally perfectly proper for a municipality to initiate land consolidation strictly on its own authority, it was politically imprudent to do so. Farmers who can vote land consolidation up or down are morally committed to it once the majority has spoken. Even when, as at Buzet, a sizable number of farmers balked as *remembrement* proceeded, such opposition lacked legitimacy because it went against what a majority had voted.

In retrospect, many Cadours farmers who attended the April meeting also felt that "they learned about the roses, but were not told about the thorns." They were given to understand that the official procedure upon which Cadours was about to embark was infinitely supple, that individual objections at every stage would be weighed by an informal subcommission made up entirely of local farmers, that they would be scrutinized by an official communal commission on which local farmers were also conspicuously represented and, if complaints remained, that they could be appealed to a departmental *remembrement* commission on which farmers also played an important role.

There were other errors. In 1959, the mayor's slate had won by only

six to five, although it is true that the minority loyally rallied to Dr. Suchich. By the time of the next election in 1965—after the demand for land
consolidation had been filed, but before it had been acted upon—Dr.
Suchich's list had won hands down with three-quarters of the local bal-
lots. Undoubtedly, the mayor had a strong political hand to play, yet it
would nonetheless have made sense to depoliticize the process of land
consolidation as much as possible. To propose at the head of the list for
the communal land consolidation commission three men who either sat
as municipal councillors or were close relatives of councillors was not astute. To have Dr. Suchich chair the subcommission that dealt with the
day-by-day quandaries of *remembrement* was to politicize the process
even further, even though the mayor, no doubt sincerely, saw himself as
a disinterested arbiter.

It would also have helped had the mayor been able to reach some ac-
comodation with the township's contingent of North African expatriates,
particularly with Barnadien, the largest landowner, most progressive
farmer, and a man who came to be generally respected, described as "a
man like *that*" (*un homme comme ça*) at which point the speaker raises a
trembling fist, thumb pointing skyward. Granted that when municipal
elections were held in 1959, Barnadien was still an isolated newcomer in
Cadours and there was no compelling reason for involving him in municipal politics. Yet in 1965, after Barnadien had become something like the
unofficial spokesman for the half dozen *pieds-noirs* who settled in Cadours after their expulsion from North Africa, Dr. Suchich did not invite
him or any other of the expatriates to join his list. Did ideological differences—*pieds-noire* were associated with the Right, the mayor with the
Left—carry over into local politics?

However desirable in the abstract it would have been to coopt the
newcomers, the mayor's choices were severely circumscribed by the legacy of the eight-year-long Algerian War that ended only in 1962. In contrast to earlier colonial wars fought by professional soldiers, the French
soldiers sent to Algeria were primarily first-time draftees and recalled
reservists, most of whom spent twenty-six to twenty-eight months devoted, among other things, to patrolling the estates of French settlers.
Exposed for the first time to the realities of a colonialist society, French
peasants' sons were genuinely shocked by its racism and injustice. Above
all they came to resent the arrogance of the rural *pieds-noirs* whose
landed properties they guarded, families who forbade their daughters to
go out with enlisted men, who, to cite the most durable myth of the
Algerian War, *nous ont fait payer le verre d'eau* (charged us for every
glass of water). If the returning veterans came back loathing the Algerian
settlers, the expelled *pieds-noirs* who scattered throughout Mediterranean France were generally embittered and often defiant, like the new-

comer to Cadours who painted black feet on the gate posts of his farm. Even had all *pieds-noirs* been models of tact, they would still have been resented: their land purchases, by driving up the price of farm real estate to two or three times of what it had been, had infuriated all those local farmers who had hoped to expand; the former French settlers alone enjoyed generous terms of credit at a time when local returning veterans were unable to buy farm land with the help of the Crédit Agricole. Whatever common sense might have dictated, *pieds-noirs* were too unpopular as a group to be given representation either on the municipal council or on one of the *remembrement* commissions. The mayor's hands were tied.

It probably would have been more feasible to give the *pieds-noirs* "credit" for whatever pioneering work they had done on their own farms that related to *remembrement*. Someone like Barnadien, unaided and unsubsidized, had, after all, managed to consolidate his farm at his own expense. It was predictable that he and others like him would be unhappy at having to pay for land consolidation a second time by forcibly sharing the communal expenses of *remembrement*, when they themselves had had to defray similar expenses unaided.

There were any number of ways in which this could have been finessed: simply exempting those who had achieved consolidation by their own efforts; or exempting them from the perimeter to be consolidated; or including their earlier expenses that they could document in the general works budget associated with official land consolidation. Admittedly, some of these ploys were of questionable legality, but considering traditional French skills at circumventing bureaucratic rules, had there been a will, there would have been a way.

Finally, the mayor would have been wise to be especially sensitive to requests linked to the *remembrement* made by North African expatriates, such as Perrot. He was not. As it turned out, appeasement was not even attempted. The *pieds-noirs*, better educated, touchier, yet more self-possessed than most local farmers, felt officially ignored, bypassed, and thwarted. They were therefore primed to take the lead in any nascent opposition movement, which in its way would for the first time permit their partial integration into the local community. To be elected president and vice-president of an Association in Defense of the Interests of Landowners Opposed to *Remembrement* at Cadours for the first time gave them some local standing.

It was even more vital that the process of reordering the field system of Cadours, that is, the specific decisions made by the communal commission and subcommission, appear responsive, evenhanded, and in accordance with the law. In this respect, *remembrement* at Cadours began disastrously with the "Inquiry Concerning the Perimeter," that is, with the two-week period in June 1966 during which landowners could formulate objections to being included in the area to be consolidated or demand that land consolidation be extended to fields that they owned in adjoining *communes*. During this two-week pause, nineteen landowners wrote out specific objections. All nineteen objections were denied. Some of the complaints did lack substance; others had to do with land owned near the village or along the main road and therefore thought suitable for building; several, including Barnadien and Perrot, asked to be exempted because they had already completed individual land consolidation on their own initiative and at their own expense. Perrot, as an alternative, proposed that if his land were nonetheless to be included in the project, consolidation should also apply to 7 hectares he owned in the adjoining township, where his land was mixed in among fields owned by two other Cadourciens, his neighbors. The formal procedure of *remembrement* authorized such extensions, but this too was turned down.

To top it off, a twentieth request was added after the deadline had expired. It too involved exempting land suitable as a building lot, only this time the request was granted. The lucky owner of the land in question happened to be a municipal councillor and deputy mayor, the would-be purchaser, the municipal secretary. Whatever the special circumstances that might have justified this decision, politically it was a catastrophe. After an impressive show of rigidity that must have put off more than the petitioners directly concerned, the communal land consolidation commission, and the mayor who was seen as guiding it, had managed to offend the powerful French streak of legalism and the equally strong revulsion against special privilege.

Another altogether avoidable error had to do with the surveyor's decision, sanctioned by the communal land consolidation commission, setting up but a single category of farmland in the township of Cadours. In classifying farmland in preparation for future reallocation, no distinction was to be made between land currently in fields, in meadows, and in vineyards. This meant that in exchange for his vineyard that might be at some distance from his home, a farmer might be given an ordinary field closer at hand. Yet in the real world vineyards required light, well-drained soil that was scarce in the township: the loss of one's vineyard was therefore bitterly resented. Similar objections were voiced when it came to trading meadows for fields. Natural meadows needed the moisture of bottom land or of the proximity of a water course. Where they were used for pasture rather than haying, they also called for shade trees or tall hedges to shelter ruminating cattle from the direct sun. Since in the 1960s everyone at Cadours kept cattle, anxiety over being awarded land unsuitable for pasturing was a legitimate grievance. Farmers' interests were apparently being sacrificed to the convenience of the surveyors.

There was also widespread skepticism about the impartiality of the subcommission assigning a varying number of points to each and every one of the 1,887 distinct plots listed in the Cadours cadastre. Each assessment was supposed to reflect not only the acreage of the field involved, but also its agricultural potential. There was widespread belief that it was all a matter of those in the know taking advantage of the trusting and the uninformed. Members of the subcommission were rumored to help themselves and their friends at the expense of those who failed to show up at hearings, primarily because they were old, poorly educated, and did not grasp the procedure. Even ten years after land consolidation had been completed, local people still trotted out their repertoire of horror stories, most of the "horror" amounting to the loss of less than half a hectare. The trick, I was told, was to overvalue your own land and undervalue your neighbor's. If subsequently exchanges were ordered, the owner of the overvalued land had to give up less than the proprietor of the undervalued fields, for exchanges were based on the points awarded and not simply on the basis of acreage. To what extent this really took place is hard to say. What is perhaps more important is that at Cadours such charges, which are raised in connection with every *remembrement*, were widely believed by a majority of the local farmers.

Finally, the particular surveyor chosen by the Direction départementale de l'agriculture (DDA) turned out to be a liability in life as in death. From the start local farmers were miffed because the regional surveyor usually employed at Cadours was not on the DDA's "approved" list. The stranger who was chosen was frequently absent at crucial moments when authoritative information could have allayed disquiet based on uncertainty and misapprehension. For example, when the maps proposing the new field system were first placarded in the municipal building in May 1968, neither the surveyor nor any member of his staff remained behind to explain the plans. The result was pandemonium, since the mayor was in no position to justify proposals that had never been explained to him.[27] According to a good many Cadours farmers, the members of the surveyor's staff could only be distinguished by the degree of their arrogance or incompetence.

As resistance to the way land consolidation was being implemented in Cadours mounted, the surveyor came under increasing pressure from the DDA that had retained him and from the Chambre d'Agriculture, the official representative of the agricultural profession, whose officials were feeling the heat from Cadours farmers. In fact, the surveyor did respond to these pressures, becoming more amenable and open to suggestions in the fall of 1968. Local farmers began to feel that they were making some headway: new plans for the field system had been thoroughly revised in accordance with widely voiced wishes. At this point, the surveyor died of a heart attack while on a business trip to West Africa. A new supervising surveyor was brought in. Aside from the fact that he had a reputation for being stiff-necked, he was in the unhappy position of being held responsible for decisions he had never made and prodded to make changes that had supposedly been agreed to orally by his predecessor. Some of the moderate opponents of the Cadours *remembrement* naturally felt that they had to refloat their campaign for greater responsiveness to local demands.

What had happened was that the posting of the proposed field distribution at Cadours had brought discontent to a head in the summer of 1968. Fifty Cadours landowners submitted a petition demanding that a stop be put to land consolidation, a petition addressed to the head of the DDA at Toulouse. The opposition to *remembrement* consisted of four distinct, if sometimes overlapping, factions: the old who resented having their habits upset; the "short-sighted and the dumbbells" (*les bornés et les cucus*, in the picturesque phrase of one former opponent); the farmers who had already completed their land consolidation on their own and who resented paying for that of the others; and, finally, a number of farmers who were convinced that land consolidation was indeed needed at Cadours, but that it was being carried out in an unacceptable manner.

The opposition petition was denied on the grounds that, once officially mandated by the prefect, *remembrement* could not legally be halted. However, the DDA did propose a meeting on neutral ground between its officials, the surveyor, and a delegation of oppositionists, with representatives of the Chambre d'Agriculture sitting in as mediators. What came out of this meeting was an official proposal that there be a delay in the proceedings while the DDA promoted dialogue between all of the parties concerned. Consequently, throughout the fall of 1968, a number of meetings were held in Cadours, allowing more airing of grievances by one side, more information on the practical and juridical implications of land consolidation on the other. The newly founded Association in Defense of the Interests of Landowners Opposed to *Remembrement* at Cadours, presided over by Barnadien, did not disband and held a series of meetings of its own.

Yet the authorities did become considerably more responsive to local public opinion, as did the communal land consolidation commission when it considered objections to the way land was being reallocated. However, this new sensitivity was not retroactive: earlier decisions on inclusion in the perimeter and on the classification of individual fields remained unchanged. Since support or opposition to *remembrement* had become tantamount to being for or against the mayor, to defuse the situation, Dr. Suchich wisely decided to absent himself from further meetings of the communal subcommission.

In the end, land consolidation ended up as a very qualified success. It is true that the number of fields listed in the land register was drastically reduced, from 1,887 to 346, and their average size increased from one-half hectare to 3 hectares, a reasonable size for modern motorized equipment.[28] Some of these fields, as in the case of Barnadien, had already been enlarged by their owners, a fact not always recognized in the cadastre. Presumably, the real change that *remembrement* achieved was somewhat less spectacular than what appeared on paper.

When it came to planning the public works affecting farm operations in the township—removal of hedges and ridges from the newly consolidated fields, the digging of drainage ditches, and the construction of field access roads—the DDA specialist tired of swimming against the tide of public opinion. He yielded to those who feared an excessive long-term burden of annual debt payments by proposing "public works à la carte." In each area of the township, farmers were to indicate to the DDA just how much they were willing to pay and plans would be tailored to their budget. The net result was that at Cadours, as at Buzet-sur-Tarn, farmers paid little and got less: for instance, drainage ditches, which would have benefited local agriculture, were largely neglected because of the expenses involved, despite the fact that 80 percent of such outlays were defrayed by the authorities. Similarly, in contrast to what took place at Saint Cézert, farmers at Cadours were unwilling to tax themselves to asphalt access roads and farmyards, improvements that would have made daily life much easier. Here, too, they threw away generous official subsidies. On the other hand, in contrast to Buzet and Saint Cézert, at Cadours more attention was paid to the ecological effects of proposed changes, so that there was no brutal leveling of all hedges and windbreaks. In any event, the grand finale did not come until late in 1971 when the music stopped. In this rural game of musical chairs, every farmer seated himself on the land that *remembrement* had assigned to him.

More than ten years after these events, most of the original supporters and opponents of land consolidation have long since come to terms with it. Moderate opponents have gone on to other things. Pascaglia, who approved the principle but not the implementation, was willing to admit that in the long run, even an inequitable land consolidation was better than none. Léon Barthez still regretted his old vineyard that he had never been able to replace because his new farm lacked the required patch of well-drained soil. He also missed the free firewood once furnished by a grove that was no longer his and that, anyhow, his neighbor hastened to cut down when the plot was assigned to him. Despite these and other reservations, the Barthez holding was now compact and therefore easier to work, as he reluctantly admitted.

In any case, with the passage of time, the Cadours graveyard has

13. Cadours before land consolidation

14. Cadours after land consolidation

hushed some of the most vocal opponents of *remembrement*, including the doughty Villeneuve, who had organized the petition of 1968 at the age of ninety-four. Barnadien, the former president of the Association in Defense of Landowners Opposed to *Remembrement* at Cadours, could still muster indignation while reliving what he considered a tale of chicanery and bad faith, but, being a sensible man, he knew that the past was inevitably receding. Only the association's former vice-president, Perrot, was still deeply enmeshed in the juridical postscript to *remembrement*. His life had come to center around endless procedural challenges to what he considered the injustices inflicted on him by land consolidation at Cadours, challenges that he had taken all the way to the highest competent authorities in Paris. For Perrot, eyes shining as he explained the intricacies of his appeals before that Supreme Court of the French bureaucracy, the Council of State, the story of *remembrement* at Cadours had not been heard out yet. There was still this appeal pending and that reconsideration by the administrative law judge required by the latest ruling of the Council of State, not to mention the civil suit against the municipality of Cadours still in the offing.

Even before the last bulldozers had left the township of Cadours, its mayor was making new plans. He was still—or again—aiming for the local lake that would make irrigation possible in his *commune*. For ten years he had been scouting locations for such a lake. By 1971, he could look to a nearby example, the Lac de Saint Cricq, 15 miles from Cadours in the *département* of Gers. In that instance, the Compagnie d'aménagement des Côteaux de Gascogne had acted as contractors, but it was the European Economic Community (EEC) that had footed the bill, Gers having been able to take advantage of a provision for aid to underdeveloped areas within the community. What made this nearby example so attractive to Dr. Suchich was not the easy financing. Apparently there was no hope of extracting further largess from the EEC for similar purposes. Suchich was most impressed by the fact that such a lake could be made to serve a dual purpose, agriculture *and* tourism. As a tourist attraction, a sizable lake in Cadours would serve not merely to enhance farm incomes through irrigation, but also infuse new life into the community by bringing vacationers and more permanent summer residents, perhaps even transforming the village into a minor summer resort. The mayor's only regret was that the land consolidation hassle had wasted so many years. Otherwise, by now people would be extolling the great lake at Cadours rather than envying that of Saint Cricq.

The mayor had learned from the experience of *remembrement*. To be sure, he negotiated extensively with the Compagnie d'aménagement des Côteaux de Gascogne on possible sites before going public, but he could justify such reticence: after all, he needed an authoritative estimate of the costs involved before presenting the project to local farmers. His first plan called for a relatively small lake, to be located at the northern edge of Cadours, the very area where Barnadien and several fellow members of the anti–land consolidation league had their farms. Dr. Suchich decided to see Barnadien to gain his support. Barnadien was already in his late forties by then and without an heir. He had no interest in launching into irrigation at this stage of his career. Moreover, in the course of the visit, he and Suchich had became involved in an inevitable wrangle over the "original sin" of *remembrement* at Cadours. Barnadien, having lost the first round over land consolidation, was not about to concede the bout: he was bound and determined to defeat "Suchich's lake." Suchich left sufficiently shaken to give up on the site he had first proposed.

When a meeting of Cadours farmers was convened in November 1971, the presentation by the representative of the Côteaux de Gascogne was straightforward, detailing exactly what advantages and costs would be involved in a lake large enough to provide both irrigation and recreation. Pascaglia, who, independent of Suchich, was also inspired by North Italian irrigation, having just finished digging his private lake on his own little watershed, was sick at discovering that he had perhaps spent Fr 25,000 for nothing. Yet he felt that public reaction to the plan for the grand, communal lake was at best lukewarm, that local farmers were leery of taking on anything so ambitious.

It was not just the conservative local mind-set, that famous *mentalité* that was at fault, although in 1971 irrigation was sufficiently novel in the region that skeptics could argue about its unpredictable future. At least two other factors also played major roles. In the first place, by comparison with the introduction of irrigation at Buzet or Saint Cézert, the proposed lake at Cadours called for a much longer-term commitment. The Compagnie d'aménagement des Côteaux de Gascogne would only underwrite the project if local farmers formed an irrigation association. That association, in turn, would contract to pay for a given volume of water for twenty-five years, with each member allocated his share of water and cost on the basis of acreage to be irrigated. Even if the summer turned out to be rainy, members would be charged for the water, whether they used it or not. Given the advanced average age of local farmers, the prospect was unattractive not merely because many were reluctant to tote sprinklers three times a day, but also because, in the absence of successors, they would not make commitments that might jeopardize the retirement to which they looked forward. In the second place, the experiences of the local equipment cooperative and, in a different context, of the *remembrement* itself, were not such as to encourage another communal effort. As few as four cooperators were eligible for government subsidies to construct their own irrigation pond over which they would retain full control, a formula that seemed much more attractive. Even if the controversy over land consolidation had never divided the farming community in Cadours, it was not certain at all that a proposal that demanded even more community involvement than *remembrement* would have been supported.

Despite local hesitations, over the next two years plans became ever more ambitious. From a scheme for a lake confined to Cadours, a plan for a much larger lake to be sponsored by half a dozen *communes* emerged by 1973. This new lake was to be almost half a square mile in acreage, with a water storage capacity of some 10 million cubic feet, larger than the competition at Saint Cricq. There were endless meetings of the mayors and groups of municipal councillors with the cantonal general councillor, with civil engineers speaking for the prefect and with hydraulic specialists from the Compagnie d'aménagement des Côteaux de Gascogne. There were interminable palavers over the cost of moving two departmental roads that would be flooded once the Margesteau was dammed up as planned. And there were the additional problems of finding a new farm for at least one, and possibly two, tenant farmers whose land would be drowned.

Above all, once again there were mass petitions to the powers-that-be and the organization of a defense association by all those, Barnadien conspicuously among them, who opposed the lake. The opponents this time rallied to the defense of the two vocal and unhappy tenant farmers whose lives were going to be disrupted once the valley of the Margesteau was dammed up. The political balance began to tip against the whole project. Could the opponents of *remembrement* take credit for killing "Suchich's lake"? Did their petitions decide the canton's general councillor to settle the hash of the upstart mayor of Cadours once and for all? Was Dr. Suchich's pet project sacrificed to a long-planned political vendetta that had merely awaited an auspicious catalyst? Was the support from the other townships only tentative all along, so that rising costs gradually transformed unease to outright opposition? Did the prefect veto the intercommunal lake because he wanted to avoid another politically damaging local free-for-all? Did the Compagnie d'aménagement des Côteaux de Gascogne pull out the plug because "Suchich's lake" competed with "its own" Lac de Saint Cricq nearby? Everyone in Cadours still had his or her favorite theory or theories, none of them readily verifiable. The only thing certain was that by the spring of 1974 the intercommunal lake was dead.[29] Later, at least two of the *communes* that had participated in the consortium, Garac and Brignemont, were to go ahead and construct communal irrigation ponds of their own, but not Cadours. The public funds already allocated to the intercommunal project were transferred to a similar irrigation and recreation plan in another part of Haute-Garonne. Not long afterwards, Mayor Suchich resigned his office.

One reason why Cadours farmers could afford to pass up irrigation was that the market for what had been one of their minor crops, garlic, began to boom in the early 1960s. Some garlic had been grown as long as people could remember. Agricultural statistics that went back to the pre–World War I era confirmed that 10 hectares of local farmland were devoted to this form of gardening, which is what garlic growing really was.[30] The violet-colored regional variety was not only distinctive, but, having a longer period of dormancy, its preservation was assured: Cadours garlic was by definition premium grade.

Traditionally, a prudent family might have devoted a tenth of a hectare of land to growing garlic. On one hand, garlic was a very risky crop, sensitive to the vagaries of rainfall—drought and excessive rainfall were equally disastrous—and to numerous, untreatable blights. The crop also required an inordinate amount of labor, as it had to be planted by hand, cultivated by hand, and harvested by hand. It was then dried in shed or barn, before being hand peeled and prepared for sale by stringing or braiding. This final preparation was a task that in late summer normally fell on the older generation on a farm. So much for the negative. On the positive side, the economic importance of garlic lay in that it was grown mainly for cash, rather than for home consumption. Garlic served as a modest savings account in traditional Cadours. Five or ten kilograms' worth of braids sold on the yearly garlic fair held every fourth Wednesday of September counted for something in the family budget of peasants like the Barthezes. Saint Antonin remembered carrying three or four chains of garlic draped over a stick to the ordinary weekly markets.

In the early 1960s local and regional garlic production began to expand tremendously, so much so that by 1967, garlic markets had become a weekly affair in Cadours for six months out of the year, providing as much a social as an economic focus for the community and the whole canton.[31] Garlic growing became generalized and individual acreage rose from a fraction of a hectare to 1 or 2 hectares on most farms. Instead of bringing a few chains of garlic to market as in his youth, in 1982 Saint Antonin had delivered as much as 400 to 500 kilograms on a single market day. That year his crop had amounted to 6 tons on about 1 hectare of land, with gross receipts of over Fr 90,000. An outlay of Fr 15,000 for cloves of seed garlic and perhaps Fr 5,000 for fertilizer reduced his net income to Fr 70,000, roughly what a French schoolteacher brought home in a year. To be sure, garlic prices were exceptionally high in 1982, but after all, Saint Antonin, like other farmers, had other crops to sell besides garlic. Garlic had indeed brought a new prosperity to Cadours farmers and one that even extended to farmers on tiny farms. At the very least, garlic had become a much bigger savings account than formerly, one that could now finance a new car or an up-to-date, though secondhand, tractor.

As to what accounted for the upswing in the demand for garlic, no one had a firm answer. For most farmers, it was just one of those things, much like the weather. Agricultural extension agents talked vaguely of some I. G. Farben plant that used garlic for a fiberglass bonding agent in Germany, of the latest fad for garlic-based cookery in the United States, or of the insatiable British appetite for condiments to spice up those ghastly boiled meats on which they were said to feed. Everyone speculated, everyone recognized that things had changed, but no one really knew why.

What made the expansion in garlic production practical was much clearer and it had to do with two technical developments. In the first place, equipment for planting garlic mechanically became available in the late 1960s, replacing the slow and back-breaking stoop labor of planting cloves by hand. By 1982 even the Barthez household, conservative as ever, had finally invested in a secondhand garlic planter, the last garlic farmers in Cadours to do so. Secondly, by the early 1970s, INRESA (Institut de recherche scientifique agricole, that is, the Institute for Scientific Agricultural Research) took some of Cadours' violet garlic and proceeded to develop a disease-free strain. Once specialized growers had multiplied the new variety, the traditional risks of garlic growing dropped sharply, though the crop could still be desiccated by drought or brought to bursting by a wet growing season. By 1983, the advent of an effective mechanical garlic harvester seemed just around the corner. Abadie had tried using such a machine in 1982, but reluctantly concluded that it would take another year or two for the manufacturer to get the bugs out of the new equipment.

Where no breakthrough seemed possible was in preparing garlic for sale. It was all very well for Colette Abadie to hire young people to help prepare her garlic for bagging as long as garlic was fetching Fr 15 to Fr 18 per kilogram wholesale as it did during the 1982 season. Other farmers insisted that when garlic fell to Fr 8 per kilogram, that is, to more nearly its run-of-the-mill price, hiring outside help soaked up most of the profit. They gleefully retailed tales of *pieds-noire* and other outsiders who were said to have planted 10, 15, or even 30 hectares of garlic, only to lose their shirt when they discovered that they could not find enough help to harvest and peel their crop. By and large, Cadours garlic remained a family enterprise in which the older generation continued to play an essential role.

It was the younger generation that was pushing for greater acreage, some older farmers complained, but the same young people shunted the hard, painstaking work of preparing the garlic for market unto others. The older generation was convinced that creating some sort of cooperative—there was already a garlic producers' cooperative in the canton, though it had attracted few adherents from Cadours proper—that would miraculously machine-process garlic in its own plant was a will-o'-the-wisp. They claimed that the only thing such a cooperative would accomplish would be to process the profit right out of the garlic.

Some of the older generation challenged the productivism of the young farmers by calling for quality over quantity when it came to local garlic. There was no substitute for intensive effort and skill, Merle maintained. He was fifty-seven years old and as a farmer had done very well for himself on only 20 hectares. He insisted proudly that he was living better than his son, the *lycée* professor, or his daughter, the accountant in the Inspectorate of Finances. Merle grew only half a hectare of garlic, but in 1982 at the Cantonal Agricultural Fair, his garlic had won three gold medals. This in turn had allowed him to sell most of his crop as seed at between Fr 20 and Fr 23 per kilogram. His wife was famous for the way she could braid garlic, which brought in those customers interested only in quality. There was one Dutchman who had driven all the way from Royan on the Atlantic coast to buy their garlic. French producers who neglected the way in which they presented their garlic were in instant competition with all the second-rate garlic producers all over the globe, from Mexico to Italy, from California to Spain. Yet beautifully presented, Cadours quality garlic had no competitor in the whole wide world. When Dutch, German, and American buyers, who knew quality when they saw it, took a look at the kind of garlic that *he* produced, they did not haggle over price.

Merle, like other local farmers and agricultural technicians, was well aware that garlic was caught up in a vast international and intercontinental market in which a drought in southern Spain drove up the price Cadours growers received for their product. Farmers shrugged ruefully, as they deplored that their own good fortune should depend on someone else's bad luck. They had also learned to adjust to the yearly winter influx of Argentinean garlic, produced in a land where the seasons were out of whack, by selling off their own before the Argentineans perennially flooded the market. Some Cadours producers would even tell you that when the Spanish growers ousted the Californians from the Brazilian market, that too affected prices in Cadours: suddenly the Spaniards did better shipping their garlic across the South Atlantic than expediting it across the Pyrenees. And Cadours farmers were only too keenly aware that garlic was a fickle market. Unlike reliable crops like wheat, corn, sunflowers, and soybeans, for which the European Economic Community set yearly minimum prices backed by import controls, garlic needed no passport: its prices shot up and down according to the whims of world-wide supply and demand.

This is why most Cadours farmers were skeptical about the future of garlic, all the while increasing the acreage that they devoted to the crop. They agreed that this bonanza was too good to last, while resenting the garlic producers' cooperative's warnings against indiscipline among growers. Only Laborie was entirely consistent. A longtime garlic grower himself, he was a veteran observer of other garlic growers. What with every farmer increasing his acreage, Laborie had reached the conclusion that a collapse of the garlic market was inevitable. "*Il faut être logique*," he maintained, "You have to face facts." And logical he was: he would plant no garlic in 1983.

Parallel to the growing importance of garlic in the local economy was the expansion of poultry raising. Just as in the case of garlic, raising fowl had been a traditional part of Cadours' polyculture. Barthez remembered his mother carrying a few chickens or ducks to the weekly Cadours market. Despite the new poultry vendors' hall built by the municipality, in recent years the Cadours poultry market had been pretty well eclipsed by larger markets in Gers that happened to meet on the same day of the week. This is where Barthez and his sister took their ducks nowadays. On their 12 hectare farm, they continue to hew as closely to the old ways as any farmers in Cadours.

As in the case of garlic, what was new for most farmers was the scale of fowl raising. The pioneers in producing quantity were Merle and his wife, long aided by their parents, for whom raising as many as five thousand fowl a year and retailing them on the farmers' markets of the Toulouse suburbs had provided much of their income. In time the parents died, garlic prospered, and they themselves began to feel their years. While they abandoned their profitable poultry operation, others took up where they had left off. The Labories had done what the Merles had started: they too sold on the Toulouse markets, though with a somewhat more modest poultry production. They were forced to quit when the wife's health declined, but their son had recently revived the family enterprise by contacting firms in Toulouse and taking orders from and delivering to their employees. After less than four months, young Laborie was selling 100 to 150 roasting geese per week.

Rising French living standards had also enlarged the market for *foie gras*, the enlarged livers of force-fed ducks and geese. Right next door, the *département* of Gers had long specialized in producing this expensive delicacy, but the number of Toulousains willing to shell out Fr 200 a kilogram for duck liver and Fr 300 for goose liver had increased sufficiently to draw new areas like Cadours into force-feeding fowl, an operation that had become partly mechanized. We have already met Mme Abadie stuffing her eight hundred ducks and geese. Stuffing two or three hundred had become common locally, where twenty years earlier fifteen or twenty would have been the norm. The days when customers bought their live force-fed duck or goose, gambling on the size of the precious organ, were drawing to a close too. The new producers sold liver and fowl separately, for the regional taste for duck and goose preserved in its own fat insured a market for both. Typically, to maximize their selling price, the poultry raisers tried to bypass rural regional markets and wholesalers. They did their best to encourage private customers to come to the farm, and, failing that, to sell directly on the farmers' markets in and around Toulouse.

Garlic and poultry were also instrumental in the rapid decline of dairying and cattle raising in Cadours, which, as in Buzet-sur-Tarn, had become universal once the introduction of the tractor had made work animals obsolete. What garlic, ducks, and cows had in common was that they were all three labor-intensive: it was no coincidence that the largest and most successful dairyman in the township, Pascaglia, grew no garlic and raised no fowl. It was also no coincidence that Pascaglia alone intended to keep his cows, though his son seemed less committed. Just as cows enhanced a farmers' income per hectare, so did garlic or poultry, particularly if the latter were fed with one's own corn. At the same time garlic and force-feeding, being more seasonal, were less demanding than the never-ending and unrelieved enslavement to a dairy herd. Goslings and ducklings were hatched or bought in June, stuffed from November on and by February the year's flock had been liquidated. Only three or four months of intense efforts were involved, compared to the daily, year-round milking that tied farmers down to their cows.

What made the switch to fowl even more attractive was that, increasingly, small dairy farmers had to confront the obvious: their herds were too small to be profitable and their obsolete barns precluded major expansion short of heavy investments. This was the argument that Paul Renal was using on his father: to build a modern cowbarn or open stable facilities took at least Fr 200,000 and ten years to amortize; the basic equipment for commercially force-feeding ducks or geese cost Fr 10,000 to 20,000. It was obvious that economic rationality was about to doom the Renals' eleven head of cattle. Within two or three years, only the traditionalist Barthez's seven cows and the modernist Pascaglia's sixty may be the lone survivors of what had been Cadours' seven hundred to eight hundred head of cattle twenty or twenty-five years earlier.

If Pascaglia, a thoughtful and level-headed man, had misgivings about this trend, it was not just a matter of sentimentality or even because his highly productive dairy herd continued to pay its way. What concerned him was the agricultural future of his own farm and of what had come to be his community. Cattle raising called for an elaborate crop rotation—in Pascaglia's own case it currently ran to eleven years—that included four years of *luzerne*, a leguminous plant, and two years of Italian rye-grass, both fodder crops that add nitrogen from the atmosphere to the soil and the roots of which also renew its humus. Moreover, land on which *luzerne* had grown retained a soil structure that made plowing easy, even at the end of a summer of drought such as that of 1982. Though other farmers' crop cycles might depart from Pascaglia's, owning cattle mandated the same crucial fodder crops. At the same time, the availability of ample manure made possible a constant renewal of humus.

The end of local cattle raising meant a cutting out of fodder legumes and grasses from the crop rotation. Plowing under corn and grain stalks, as the agricultural extension agents had recommended, seemingly did not provide adequate raw materials for new humus in soils that were no longer manured. Farmers were digging up undecayed stalks three and four years old. The lack of manuring also meant that the chemical fertilizers had to be increased and in the long run the chemicals applied acidified the soil, a trend that visibly affected crop yields. The problem was not irreversible, but the solution—very heavy liming—was expensive, as Pascaglia discovered. Besides, the loss of humus was magnified by the abuse inflicted by the powerful new generation of tractors that had become the norm among the larger farmers of the township. Tractors had at first permitted utilizing virgin topsoil that cattle-drawn plows had never turned over. Yet in recent years, farmers with their new 100 and 130 horsepower tractors had been plowing so deeply that they were diluting their humus with infertile clay dredged up from the bottom.

This new problem added to the increasing threat of erosion that motorization in this hill country had posed from the start, but which the new equipment had aggravated. Before the coming of the tractor, contour plowing had been the rule in the Cadours hills. This did not reflect some special ecological sensitivity or an uncanny folk wisdom, but merely recognized the stubborn fact that cattle did not have the physical strength to pull a plow uphill. Tractors did have the power to do just that and motorized plowing up and down was obviously safer than plowing with the tractor tipped sideways on an incline. Besides, once tractors were improved to carry, rather than drag, their plows, plowing crosswise had become impractical altogether—or so farmers insisted. In the last ten years, the introduction of the rotative harrows that, willy-nilly, moved earth down-slope as their moving teeth chewed up the clods, had completed this arsenal of creeping destruction. Cadours' soil, the regionally famous *terrefort*, invariably required plowing before the onset of winter and nowadays more than before, such plowing invited erosion. All winter long, the land is pounded by intermittent rains: one night's downpour may wash 10 inches of topsoil to the bottom of a slope. Pascaglia was not worried about tomorrow. Even if his son sold the dairy herd upon taking over, the land the father had nurtured and enriched was likely to retain its fertility and soil structure for as long as fifteen years. Yet beyond that, the father feared, loomed a diminished future.

That diminished future also had its human dimension. As at Saint Cézert and Juzet d'Izaut, local people shook their heads when the question of the agricultural future of their community was raised. Only about one farm out of three had an assured succession and half of all farmers were certain that no son, son-in-law, or daughter of theirs would take over the family farm. In some cases individuals clung to illusions. Merle, for instance, liked to talk about the possibility that his son, *le professeur*, who already had a weekend retreat at Cadours, would choose to juggle teaching and weekend farming. As many farmers were liable to point out, the question seemed every bit as burning twenty years earlier, yet two decades later, every patch of land in Cadours was still being farmed. The predicted doomsday of agriculture in Cadours somehow never materialized.

If the recent past is any guide to the future, childless farmers will tend to pass on their property by contract to younger farm couples, exchanging their farms for home nursing care in their old age. Farms left to city-dwelling relatives will probably be rented out, rather than sold, with the farm houses becoming rural hideaways, *résidences secondaires*. Remaining farmers will consequently enlarge their holding, even though French law, as we saw, aspires to prevent an excessive accumulation of farmland by any one farmer.[32] With labor even scarcer, equipment will have to become ever more powerful and agriculture more extensive, in the latter case reversing the trend of the last decades. Under these circumstances, farmers are likely to become even less careful of the soil that they work, justifying Pascaglia's qualms about the future. Unless the producers' cooperative happens to stumble upon cheap, mechanical means of peeling the local condiment, the famous violet garlic of Cadours will no longer be grown locally, because preparing garlic for market takes too much work. But since usually the past is a very unreliable guide to the future, perhaps Cadours will find its own original way of muddling through.

5

Where Have All the Sharecroppers Gone?

THE main highway to Castres just bypasses Loubens-Laura-gais.[1] Some twenty miles out of Toulouse, the turnoff sign on *route nationale 126* marked "Loubens, 1 km" is no more conspicuous than the handwritten sign advertising the village's one retiring restaurant. The unnumbered side road winds gently uphill past a couple of ancient farms with grazing cows, a bucolic scene with which a seventeenth-century Dutch landscape painter would have felt at home, then past a few glaringly new stucco villas glued to the hillside that form a sort of suburb to the village.

The logical place to begin your tour of Loubens is at the town hall just as you enter the village proper. There is no mistaking this official-looking two-story stone structure set back from the street for anything else. It may be a bit shabby, but you cannot miss its air of authority. Stop, open the wrought iron grille, and step inside the pebble-covered play yard. There is no point trying the big entrance in the middle: the municipal office is closed except on days when the mayor meets with his council and the part-time secretary, who teaches school in an adjoining village, keeps office hours an hour or two beforehand. The two doors on the left and the right are unlocked all right, but you would not want to disturb the children who are in class. So all you can really do is look at the municipal building from the outside. Still, the fact that school is in session at all in Loubens is hardly something to be taken for granted: keeping the school open was listed as one of the mayor's major accomplishments on his reelection campaign flyer. The second floor is occupied by the teachers' apartments and is of no interest, but if you look higher up and squint, you may just be able to make out the carved stone medallion representing Loubens' municipal crest just under the gable of the roof: a mother wolf suckling her young, a crest so eroded that it could be mistaken for the more famous symbol of the city of Rome with *its* she wolf nursing Romulus and Remus. No, Loubens makes no claim to being a second Rome, but the village name, legend had it, was derived from the French word for "she wolf." At least this is the tale that impelled the lords of Loubens to choose that particular heraldic emblem. When the French Revolution abolished lordship, the authorities saw fit to bestow their former lord's crest on the village community of Loubens.[2]

Having completed the inspection of Loubens' seat of authority, you

15. Half-timbered houses on Loubens' main street

may be ready to take in the rest of the village by continuing up the main street, the rue Tolosane, past the municipal building. The rue Tolosane is Loubens' showcase, for on the right side it is lined with half a dozen stately, half-timbered brick houses two and three stories high, of the sort that the French call *des maisons bourgeoises* to indicate residences that only solid, propertied, and well-established families could afford to build and maintain. If you have your camera, this is your chance for a good shot of a handsome facade of wood and brickwork. You don't see many better ones around, provided you don't object to getting the prosaic café-restaurant and the equally ordinary grocery into the picture.

Once you have put away your camera, you have some choices to make: taking a left at the 150-year-old market hall, a monument to the vanished Wednesday village market, you may inspect the Grande rue. Past another handsome *maison bourgeoise*, once the master's house of the Saint Jean-la-Croix estate, you end up at the post office, whose continued presence represents another notable rearguard victory for the mayor. If you turn back, you can go another short block and then saunter down the rue Vaurese that runs parallel to the Grande rue. On second thought, since there is nothing whatever to see there, you might as well keep going straight on rue Tolosane that leads to the sixteenth-century Gothic church, past the duck pond at the square, to the dark, ominous stone walls of the château and its bleak western facade. From what little one can see from the square, the chateau is Middle Ages going on Renaissance. The impression it conveys is not so much that of dilapidation, but of centuries of accumulated grime. The Gothic church is undistinguished inside and out, having been patched up so many times that it even lacks the patina of old age. If you continue on past the church square and the monument to the war dead, you are practically out of the village, though on your left the wall surrounding the château's neglected grounds provides a sort of continuation. By and by, this would take you to the cemetery that is just under the wall. When the weather is good, you might find a few old men playing *boules* not far from its entrance.

But you might as well retrace your steps. Along the way back into the village, you come to an inconspicuous fork that you are likely to have overlooked on the way out: take the left street descending to the two ruined windmills; you are in the faubourg Al Barry, the south end of the village, where the day laborers and other poor villagers used to live back in the eighteenth and nineteenth centuries. Don't miss the panoramic view of the countryside before you wander back to the bistro for refreshments.

Well, that's it. You have seen what there is to see. All you need to complement your grand tour of Loubens are a few vital statistics: this is a *commune*, which, at its peak in the 1840s, was home to some 750 people, about half of them living in the village. The current population for the village proper stands at about one hundred; if you include the two and a half square miles of surrounding farmland that is also part of the township, nowadays the total population adds up to less than two hundred people.

Indeed, for the most part, the demographic history of Loubens tells a by now familiar story. Compared to some of the other villages that we have examined, Loubens was already a very densely populated community in the eighteenth century. At the time that the land register of 1731 (the earliest to have survived) was drawn up, Loubens may already have had a population of some 600 people.[3] By 1790, when the first full-fledged census was taken, the community claimed 690 inhabitants. The population peaked between 1825 and 1855, when it hovered around 250—300 people per square mile, an amazing density for a purely agricultural community capable of producing only one crop a year. Thereafter, unevenly but steadily, the *commune* has lost population at an average rate of about 10 percent per decade, a decline that may recently have been halted by the reverse migration of Toulousains in search of country living. This meant that, as in other communities within reach of a metropolitan area, the most recent official figure, 189 inhabitants, was somewhat inflated by including the owners of weekend and holiday homes, not to mention a few newcomers without family links to the community, who, as in Buzet, commute to Toulouse.[4]

To account for the dynamics of this population decline is more complicated than to trace its statistics. It is likely that, as elsewhere, the landless
and land-poor furnished the first contingents of emigrants and, well into
the twentieth century, continued to lead the exodus.[5] In 1886, the village
school teacher, who witnessed the first twenty or twenty-five years of this
rural emigration, put the blame on the false glitter of the city, but high
wages offered in the prospering commercial vineyards of lower Languedoc may also have attracted some of Loubens' landless poor.[6] In contrast
to Saint Cézert, where the population began to shrink when farm laborers were displaced by the first wave of mechanical farm implements, at
Loubens there is no evidence that the early emigrants were pushed out.

Another smaller, but pivotal, group that had largely decamped by the
turn of the century were middle-class families who were "bourgeois" in
the original French meaning of the term in that they lived from the fruits
of their property without exercising a profession: between 1836 and 1896,
there was a striking decline in the number of residents listed simply as
"proprietors" on the local census lists. From the time of the French Revolution to the middle of the nineteenth century, this was not only the
dominant group of major taxpayers, but also the class from which a good
many municipal councillors were drawn. Why this rural bourgeoisie, already solidly entrenched in the Loubens of the 1730s, should have given
up on country living in the course of the nineteenth century remains a
puzzle. Most of the twelve or fifteen personal servants in their employ
seem also to have left with their masters and mistresses.

From the turn of the twentieth century on, artisans and grown children of artisans constituted a new contingent of emigrants swelling the
continuing outflow of farm laborers. A few, such as tailors and seamstresses, may have depended on middle-class customers who were becoming increasingly scarce. The construction during the first decade of
the twentieth century of a narrow-gauge railroad linking Loubens both to
Toulouse and to the cantonal seat, Caraman, may have undercut others
by putting them in direct competition with craftsmen and shopkeepers
in Caraman and even Toulouse. For still others, the new mobility may
have revealed new opportunities elsewhere. The few families of railroad
workers attracted to Loubens did not offset this drain. Artisans continued
to depart: in the fifteen years between 1896 and 1911, the number of
Loubens seamstresses contracted from nine to five, of carpenters from
eight to four, of wheelwrights from three to one, of masons from seven to
three.

As the population continued to diminish, between 1911 and 1926 most
of the professionals residing in Loubens disappeared. They died and
were not replaced, as in the case of the surveyor; or else they picked up
and left, as the pharmacist did, who went off to manage a pharmacy in

Toulouse. By the mid-1920s Loubens had been deprived of its resident physician, its pharmacist, its notary, and its land surveyor. Their death or departure coincided with the closing of the parochial school and the removal of its staff of teaching sisters, which left the priest and the lay schoolmaster as the only professionals in the community.

During the same period, 1911 to 1926, the farm population residing in the countryside began to shrink markedly. Even though the number of farm families declined only from thirty-seven to thirty-five, the average number of people on each farm dropped so as to reduce the rural population from 185 to 141. Perhaps the general trend toward limiting family size now came to affect sharecroppers as it had earlier affected property owners; the diffusion of labor-saving plows and reapers may have made smaller families economically viable, particularly at a time when, in the 1920s, the increased bargaining power of tenants made farm stewards eager to accept almost any willing family. Anecdotal evidence also suggests increasing restiveness among young adults in the traditional patriarchal peasant household. Extended families had never been the rule at Loubens, but they now became virtually extinct.[7]

Another reflection of this general population decline may be found in the rising proportion of the old in Loubens. Young adults normally comprise the majority of emigrants. If such emigrants return at all, it is to retire and die in their native village. This pattern had already begun to show up by the end of the nineteenth century, when 17.4 percent of the population was sixty years old or over.[8] After World War I, the figure rose briefly to 30 percent, declined to as low as 19 percent at the end of World War II, but rebounded to 30 percent in 1962 and to 35 percent in 1975. Nowadays over one-third of the *commune's* total population, and nearly two-thirds of the people inhabiting the village itself, are of retirement age. Yet as we have learned from our examination of other villages in the French southwest, the decline and aging of the rural population has become a commonplace in no way unique to Loubens.

Geography accounts for some of the more characteristic aspects of Loubens-Lauragais' agricultural history. The township is representative of an important rural region that takes in the eastern portion of the *département* of Haute-Garonne, a part of western Aude, and the southern fringes of Tarn. The name "Lauragais" derives from an obscure village in the Corbières hills, Laurac-le-Grand, once a medieval stronghold whose lords dominated the plain.[9] Not that the traditional description of the Lauragais as a plain is really accurate. Within Haute-Garonne, the region rather consists of a series of wide and shallow parallel valleys created by insignificant streams, valleys separated by low ranges of gentle hills running southeast to northwest. Villages, often with names recalling the Muslim incursions of the eighth century, usually occupy the high ground.

Loubens is typical in both regards: the *commune* takes in part of the wide valleys of two brooks, the Girou and the Vendinelle, while the village itself lies on top of the only discernible hill. The two streams, until the recent decline of cattle raising, played an important role in local farming: their moisture created the only natural meadows capable of surviving summer droughts, though the streams could also inflict damaging floods during February and March. In contrast to many other *communes* in the region, aside from the south side of the knoll on which the village of Loubens itself is situated, the undulating land is rarely so steep as to preclude plowing. In the entire township of Loubens, there is only one farm, En Tempeste, that has a few acres of land where the incline makes cropping impractical.

Loubens also shares in the bounty of the excellent regional soil, the *terrefort*, that we encountered on the right bank of the Tarn in Buzet and, in a less fertile form, in Cadours as well. The advantage of *terrefort*, it may be recalled, is not only its fertility, but also its ability to store water during the summer when rain is scarce in the southwest. The constraint of *terrefort* is that it needs to be plowed before the onset of winter, even for crops that are not put in before spring. When that constraint is respected, the local soil will not guarantee the success of summer crops, but it will certainly enhance their chances compared to other soils. The Lauragais has long had the reputation of being *the* wheat country of southern France, but at least since the eighteenth century, it has also been one of its chief corn-producing areas.

The historical role of the Lauragais as grain exporter has had as much to do with the region's access to good transportation as with the fertility and water-retention of its clay soil. From the end of the seventeenth century, when the Canal Royal des Deux Mers, the great canal linking the Atlantic to the Mediterranean, was opened, the Lauragais has had cheap bulk transport at its disposal. Throughout the eighteenth and into the nineteenth century, Marseilles and the other Mediterranean seaboard cities relied on Lauragais-grown wheat shipped by canal barge to make up their own regional grain deficit.[10] Even as late as the 1860s, after a railroad was put into service along the same route, much of the wheat was still being shipped by barge.[11] Ironically, today when the Lauragais is producing more grain than ever before, the old canal caters almost exclusively to tourists renting houseboats and to transient yachtsmen.

Yet the traditional influence of Toulouse on the region has shaped the Lauragais as much as its good soil and convenient transport. On the left or western bank of the Garonne river that slices modern Toulouse in half, the soil is generally poor gravel, or, as in Saint Cézert, the sterile, pul-

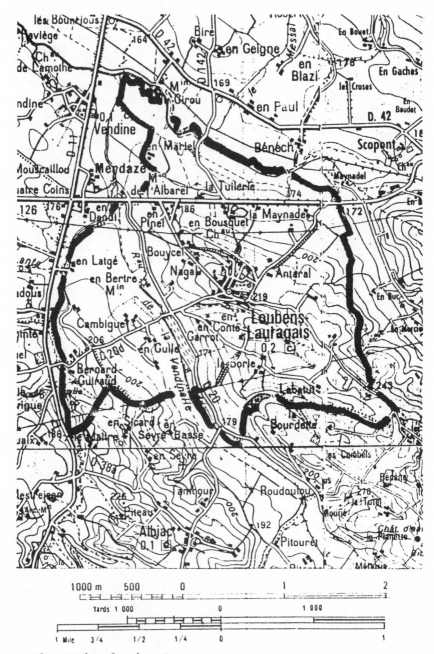

16. The township of Loubens-Lauragais

verized clay we have encountered as *boulbène*. Traditionally, those whom the city had enriched—merchants, professionals, nobles of the robe clustered around the *parlement*, the regional high court of Langue-doc located in Toulouse—chose to transform their urban wealth into landed prestige by buying estates on the best land within easy reach of the city and that meant estates in the Lauragais.[12]

Land ownership, moreover, bought more than respectability. In the sixteenth and early seventeenth centuries, such land ownership had al-lowed participation in the economic boom fueled by pastel, the dye plant that made fortunes for the Toulouse mercantile community. When the market for pastel was ruined by competition from tropically-grown in-digo, the completion of the Canal des Deux Mers permitted a reconver-sion to wheat as the regional cash crop.[13] As late as 1771, when for the sum of 130,000 livres, Joseph de Gounon, an ennobled Toulouse cloth merchant and former city alderman, purchased the lordship of Loubens and the lands that went with it from the bankrupt count of Bournazel, the estate inventory made a point of mentioning that Loubens was former pastel country, with two decaying pastel mills on the estate to prove it.[14]

In turn, this extension of the urban elite into the countryside gave a unique social configuration to the Lauragais. In examining other villages, we have encountered peasants, such as those of remote and mountainous Juzet d'Izaut, who were left in practically undisturbed possession of their overpopulated lands. We also met the yeomen farmers who dominated Saint Cézert, even while sharing their township's acreage with noble, later bourgeois, absentee landlords, and leaving a field here and a field there to a multitude of peasant small-fry. In both eighteenth-century Ca-dours and Buzet, noble and bourgeois landowners, many of them absen-tees, had owned half or more of the land. At the same time independent peasants, some working with plows, others *brassiers* turning over their land by spade, managed to hang on to the other half. In all of these com-munities, those who got their hands dirty owned a substantial, and often growing, portion of the land. This was not the characteristic pattern in the Lauragais, where independent peasants owned an insignificant share of the land, playing a marginal role in rural society. Although landowners might be aristocratic or bourgeois, local residents or absentees, their holdings ranging from a modest farm or two to large clusters of farms forming a major estate, the Lauragais was, and remained until the 1950s, preeminently landlord country.[15]

Because the social structure of Loubens revolved around landlord-tenant relationships and because contemporary agricultural modernization has turned these relationships upside down, this chapter, in contrast to the others, will focus on this major problem. We shall also be able to spell out and analyze in some detail not only how this tenant system evolved socially, but also how it developed agriculturally from the late eighteenth to the early twentieth centuries. In contrast to the other four villages that we have examined, such a detailed analysis is possible in Loubens because continuity of landownership and family traditions have preserved a good many of the landlords' and farm stewards' records. Because sharecroppers rarely struck roots in a community like Loubens, our story will rely much less than did the other chapters on the case histories of individual farmers. Loubens was a different sort of community and it is here deliberately treated in a different sort of way.

Certainly by the eighteenth century, like the rest of the Lauragais, Loubens was already landlord country, although applied to the prerevolutionary ancien régime, the term "landlord" is itself ambiguous. Legally, it harked back to feudal notions of "lordship" (*seigneurie* in French), a form of property traditionally linked to aristocratic privileges. When, for example, Joseph de Gounon, squire (*écuyer*), "Lord of Marquiset and other places," added "de Loubens" to his name, he automatically became, with Maître Bouscatel, his appointed local judge as proxy, Loubens's very own fountain of justice. By the same token, Gounon also used his traditional political authority to name, from a slate of six presented to him by village notables, the three consuls representing the three recognized status groups of the village—bourgeois, craftsmen, and peasant landowners—entitled to manage local affairs.[16] As lord, Gounon also was the monopolist who leased the butcher shop and the mill to whomever he pleased. Better than that, Gounon was also entitled to exact what amounted to yearly taxes from his subjects, fixed payments in money and in kind that added up to well over 1,000 livres, a very tidy sum.[17] As we know, the legislative reforms of the French Revolution put an end to this form of "feudal" property.

Yet the de Gounon de Loubens's wealth and power were hardly dented by the abolition of "feudal" rights (and the change of his name from Gounon de Loubens to Gounon-Loubens that went with it), because his real power did not depend on lordly privilege. The former lord retained the nine farms he had owned outright in Loubens, farms amounting to about one-fourth of the townships's acreage and including some of its most desirable land. Gounon's income as *mere* landowner was seven or eight times as great as the feudal dues that he had collected before the revolutionaries had outlawed them. In that sense of lordship, Gounon remained the richest man among his new peers.[18] And if he was no longer the *official* master of Loubens, he could still lord it over *his* sharecroppers, *his* tenants of every description, *his* servants, harvesters, and day laborers. To be sure, his position was no longer unique. Plenty of lesser landowners enjoyed the same sort of economic and social power on a smaller scale.

The dominance of landowners, who derived some or all of their income from land, but who contributed no labor and little in the way of entre-preneurship or managerial skill, was and remained deeply entrenched in Loubens. In 1790, for example, the tax rolls listed thirty-six Loubens farms that were worked by someone other than the owner or owners. In any event, some of the largest landowners were absentees, including Gounon-Loubens himself and the multiple heirs who shared his estate by the 1820s. At the end of the eighteenth century, 40.6 percent of the land in Loubens was owned by absentees. Resident bourgeois, merchants, and professionals—the only families in Loubens keeping personal ser-vants—owned another 29.8 percent. All together these dominant upper- and middle-class groups owned 70 percent of Loubens' farmland.[19]

A few village tradesmen and shopkeepers were also members of the magic circle of landlords, and there was also a handful of peasants owning their own farms. The village baker and the innkeeper each owned two farms worked by sharecroppers; one prosperous Loubens carter owned a single farm or *métairie* as they are regionally known. Yet on the average, shopkeepers and artisans would deem themselves lucky if they owned a few small fields and a bit of meadow. Similarly, although many farm workers owned smidgens of land, only six resident peasants and one from an adjoining community had holdings substantial enough to be described as farms. The unchallenged cock of the roost was one Olivier, an illiterate peasant and long-time councilman, who, as the fourth richest landowner in Loubens had a domain so large that he too had to hire sharecroppers on part of it. Yet most of the fifty local smallholders inscribed on the tax rolls averaged little more than 1 hectare a piece. These small peasants without plow teams were evidently dependent on day labor, some casual form of tenancy, or both to assure their family's survival.[20]

If anything, in the course of over two hundred years between the drawing up of the land register of 1731 and that of 1948, the eighteenth-century pattern of property holding, of farms owned by those who did not work the land themselves, was reinforced. In the eighteenth century the seventeen most highly assessed landowners in Loubens owned a little more than 70 percent of all land; three of these seventeen were working farmers.[21] By 1948, the top seventeen land taxpayers had amassed 81.3 percent of the land, but only one, the village brick and tile maker, per-sonally worked his farm of 13 hectares. What is more, by the latter date among properties in the 5 to 10 hectare category, there were only two peasant-owned and operated farms. The fifty-nine other miscellaneous owners who shared another 57 hectares, less than 8 percent of the *com-mune's* land, could certainly not count on farming to make ends meet.

The overwhelming majority of Loubens' farms were worked by tenant farmers. This was true in 1791, when that year's land tax rolls listed thirty-six such tenant farmers and, as we shall see, tax rolls were likely to understate their number. Most of these owned no land of their own whatsoever, although they paid one-half of the land tax at which their master's farm was assessed. The same pattern still prevailed almost 150 years later. In 1936 of the thirty-one wheat producers in the *commune*, only six were working on their own property, sharing a mere 8.2 percent of all local farmland.[22]

Farm tenancy at Loubens meant one of two competing systems, sharecropping, which predominated into the 1960s and still lingers on, and *maître-valetage*, popular in the mid–nineteenth century, but becoming locally extinct in the first quarter of the twentieth. Throughout most of the period since the early eighteenth century, sharecropping was the more widely prevalent of the two forms of farm tenancy. Almost any eighteenth-century sharecropping contract may serve as a handy introduction to a system that underwent only minor changes in the intervening centuries.

The most immediately striking aspect of these documents is the repetition of the phrase "half and half" (*à demi*) that runs like a litany through these contracts. Half of all seed was furnished by the landlord, to whom went half of the wheat, the main cash crop. Other grains and legumes were similarly divided. The proceeds of the vineyard were divided half and half, as were the vine clippings (*les sarments*), prized for kindling in a region short of firewood. Such firewood as there was, harvested annually or biennially when hedges were cut back, was also divided fifty-fifty, as were the cuttings from willow trees from which baskets were woven. The tenant promised to pay half the purchase price of the one or two piglets he was legally obligated to fatten each year. The landlord pledged to pay for half the bran needed to produce the fattened hog or hogs and would be awarded one-half of the proceeds from selling it or them at the yearly Loubens fair in November. Tenant and landlord shared equally the cost of the goslings bought and the sum that they fetched on the market six months later.

There were other significant provisions. Typically, an eighteenth-century contract, usually less than two pages long, began with the tenant's promise to farm faithfully and conscientiously (*en bon ménager et père de famille*) as the invariable legal formula had it. What the tenant owed in cash was also spelled out very precisely, namely a sum that amounted to half the land tax on the property (traditionally owed in full on Saint Bartholomew's Day, 24 August) and a rent paid in fowl and eggs, so many brace of chickens either at Easter or at Michaelmas (29 September), so many brace of hens on Saint John's Day (21 June), so many brace of fattened hens (*gélines*) on All Saints' Day (1 November), and so many brace of capons for Christmas, plus the 210, 250, 300, or 350 eggs per year specified by a given contract. The money value of this rent in kind seems to have been in the neighborhood of one-third of the sharecroppers' part of the land tax burden. The landlord promised to furnish and maintain not only the needed farmhouse and out buildings, but to provide the required work animals over which he retained full ownership and for which the tenant was financially responsible upon leaving the *métairie* at the expiration of the lease. Any increase in the livestock was, of course, divided half and half.[23]

Since people were familiar with local customs, there was much that contracts left unstated. Every one understood, for example, that a landlord's or a tenant's notice had to be given prior to the last day of January and that notice by either party called for the tenant's departure either by 1 November or, less commonly, 11 November. Everyone also knew that a new tenant did not participate in the first wheat harvest of the farm he had just rented on shares and that he did not pay his half of the land tax on the new farm that first year either. Instead, he received his share of the wheat crop from the farm that he had just left and in return paid his land tax there. No one needed to spell out that landlords expected their tenants to maintain what had become the traditional three-field rotation, including fallow, which the diffusion of American corn in the eighteenth century had made possible in the Lauragais. Sharecroppers, as any other peasants, were expected to prepare the land for crops or fallow by four plowings (*les quatre façons*) in the regional formula. Everyone knew, moreover, that the division of the crops took place only after the tithe collector had taken every tenth sheaf in the field, after the harvester jointly received every tenth sack of threshed grain (by the late nineteenth century this participation had risen to every eighth sack), and after as little as one-sixth to as much as one-third of the remaining wheat crop had been set aside for seed.[24] Judging from farm inventories drawn up for one landlord in the late eighteenth century, at that time it seems to have been taken for granted that it was he who furnished the necessary agricultural implements, although there may have been exceptions.[25] Only rarely did these brief contracts spell out the clauses that longer-winded twentieth-century agreements stipulated: a sharecropper was duty-bound to devote his and his family's entire efforts to working the sharecropped farm and that any work whatever on land owned by someone other than the landlord (including land owned by the sharecropper himself) subjected him to a heavy fine or to eviction. It was also normally taken for granted that the contract implied the labor of all family members active at the time of signing. If a working adult were to be disabled or to leave, even to do his obligatory military service, the sharecropping family was obligated to replace him with a suitable hired hand.

Presumably in the last years of the ancien régime any violation of the local custom would have ended up before Judge Bouscatel, who headed Gounon-Loubens's manorial court; after the French Revolution and right into the twentieth century, the justice of the peace sitting in Caraman, the cantonal seat, enforced the same customs by persuasion if possible, by fines if necessary. As late as 1907, a sharecropping contract for the farm La Maynade, drawn up by the local notary between Jules de Gounon-Loubens and the three brothers Ferrières, specified that any dispute arising would be settled according to the custom of the canton of Caraman.[26]

The sharecropping system was supplemented by a sort of subsidiary tenancy that secured the additional labor supply needed during peak periods. On the larger farms, a family of sharecroppers would be required to hire one or two subtenant families to whom the owner would, for a modest sum, rent a house located on the sharecropped property. The main duty of these subtenants was to help with the harvesting and threshing of wheat, for which they jointly received a share of the grain as their cut, known locally as *escoussure*, that was deducted from the sharecropper's own share. The sharecropper would also provide each of these families of subtenants with 1 hectare of crop land to be planted to corn and worked by spade on fifty-fifty shares with the landlord.[27] For any other work undertaken by them, the landlord hired these harvesters at the customary wages of day laborers. This arrangement meant that, at no cost whatever to himself, the landlord could count on "captive" workers just when seasonal labor was in short supply. A subsidiary advantage for the landlord was that the spade turned over the soil twice as deeply as the plow, thereby improving humus content and crop yields. By annually reassigning plots of land to be spaded, a landowner could, over a period of years, count on having his entire farm deep-spaded, and thereby improved, at no cost whatever.

If this arrangement seemed to come down rather heavily on the side of the landlord's convenience at the sharecropper's expense, such unbalance was by no means exceptional. Nor was it merely, as one sharecropper I met bitingly remarked, a matter of everything being shared equally except the work. Indeed, every variant of sharecropping was, from the eighteenth century on into the twentieth, essentially servile, with the lessor in the clear role of the master, who, personally or through his farm steward, could and did intervene continually, and not always predictably, in the sharecropper's farm operation. By the very nature of their status, sharecroppers were interchangeable, transients who filled a slot, but who were not fully individualized for the community in which they happened to farm. Though their contracts made them pay half of the land tax, it was not uncommon to find them nameless on the official tax register, entered as "the sharecropper of Monsieur Bouquiès" or "of the Château." As late as the middle of the nineteenth century, sharecroppers' children were routinely excluded from school: at least names of such children never appeared on the official municipal list of those exempted from paying school fees and this discrimination was tantamount to exclusion.[28]

If a sharecropper was never accepted as a full-fledged member of the farming community, he was equally excluded from the community of free adults. These tenants were, for example, explicitly or by custom, barred as free agents on the village market. It was the prerogative of the landlord or his steward to buy and sell, even though the sharecropper was entitled to half the receipt or obligated to defray half the costs. On the farms attached to one of the two châteaux in Loubens—by the early nineteenth century the aristocratic Gounon-Loubens acquired a bourgeois rival in the Hourlier family—it was customary to demand personal services like laundering from sharecroppers' wives and to draft their husbands as occasional grounds keepers around the manor house.[29] More generally, sharecroppers were specifically required to cart their master's crop to market or to an alternate residence, usually with the number of unremunerated trips (picturesquely called *voyages de charité*) spelled out, together with their permissible destinations (Toulouse, Villefranche, Puylaurens, Revel, Caraman) or with specifications as to what was considered the maximum range of such a trip (within 5 *lieues* or about 15 miles of Loubens was common). One very explicit and detailed early–twentieth century contract summed things up neatly: "It is agreed that the lessor (that is, the landlord) or his representative have supervision over all activities and generally direct the farm work and the operation of the farm."[30]

The obligatory rent paid in differing fowl at specified feast days punctuating the season was a medieval ritual of domination and submission far more than a purely economic squeeze. This was tacitly recognized by the two parties. In the first quarter of the twentieth century, Loubens sharecroppers waged guerrilla war against these ritualistic remittances. As late as the 1920s some of the descendants of the Gounons fought bitterly, and with success, to preserve their right to collect in kind rather than in cash.[31] Absentee landowners went to inordinate and uneconomic lengths to have the chickens that were their due shipped to their residence, evidently because eating one's *own* fowl was a potent symbol, which, nearly 150 years after the French Revolution, still bespoke of lordship. One absentee Gounon heir living in the foothills of the Pyrenees near Pau was deeply offended when *her* Loubens sharecroppers refused to stuff geese for her: her farm steward's remonstrance that Madame could have the geese stuffed closer to home at no greater cost and with less inconvenience obviously missed some important point. Her demand had nothing to do with convenience, but reaffirmed her membership in the landed classes. It was as much part of the aristocratic mystique as the continual, and equally uneconomic, exchange of "home-grown" farm produce among a circle of relatives linked by ties of blood and marriage.[32]

Where the landowner was a local resident or vacationed on his Loubens property, such affirmations of dominance and submission were likely to be more intrusive and much cruder. There are former sharecroppers still alive in Loubens who have bitter memories of just how they felt as children when, on a whim, the master had come riding across the fields on horseback to inspect his *métairie*. In their old age they could still evoke the humiliation of watching their father doff his hat, their mother hold the stirrups, and both parents being addressed by their first name and with the patronizingly familiar *tu*.

If the status of the sharecropper was that of dependency, he competed with another type of servant-tenant known as a *maître-valet*, who was one whole notch lower on the dependency scale. Unlike ordinary live-in farm servants who were normally unmarried young people and often still adolescents, *maîtres-valets*, like sharecroppers, were hired as a family. In an agricultural system geared to wheat production, the primary responsibility of the *maître-valet* family was to prepare, sow, and take care of the wheat crop in return for a fixed wage in kind of so much wheat and corn per year; in addition, provided the family also took care of the harvest and the threshing, it received the same one-eighth of the grain normally allotted to the harvesting and threshing crew. Where crops other than wheat and where animal productions were concerned, *maîtres-valets* were treated exactly like sharecroppers: they too paid their rent in fowl and eggs; save when it came to wheat production, they too shared farm outlays, profits, and losses with their master on the same basis as sharecroppers.

Since so much was similar, where did the essential differences between *maîtres-valets* and sharecroppers lie? In the first place, as the contract specified, "as long as the lease was in force, the *maître-valet* was to do what he was told by the master with regard to anything having to do with the working of the farm, the farm itself or the land."[33] In short, even more than in the case of sharecroppers, the *maître-valet* was at the beck and call of his master or of the latter's steward. Secondly, on the average, the *maître-valet* paid for his modicum of security by a standard of living lower than that of the sharecropper. The master became practically the sole beneficiary of what was the chief cash crop in the Lauragais, wheat, which, compared to other field crops, had traditionally brought in two-thirds of the income.[34]

A French scholar who studied the records of a progressive eighteenth-century estate near Saint Felix du-Lauragais (about 15 miles southeast of Loubens) has argued that hiring a *maître-valet* rather than contracting with a sharecropper was a landowner's gamble on a good wheat crop fetching high prices. If wheat ended up having a good year, the land-owner would be better off than if he had to split his profits from wheat with a sharecropper. If, on the other hand, because of unfavorable weather or a glutted market, it turned out to be a bad year for wheat, the landowner would have fared better with a sharecropper than with a *maître-valet*, because a sharecropper would have had to assume half the losses, while the *maître-valet* was guaranteed a fixed income.[35]

There is probably something to be said for this very Cartesian explanation when dealing with an enlightened eighteenth-century landlord imbued with what were then the latest principles of *laissez-faire* economics. By contrast, the sketchy evidence from Loubens suggests that most landowners did not switch back and forth between sharecroppers and *maîtres-valets* with the trigger-like response of a Chicago commodities broker. Because most landowners were conservative, tradition-minded, or preoccupied with other business—for instance, down into the 1820s, the first two generations of Gounon-Loubens may have continued to make more money as cloth merchants than as landlords—their response to changing economic opportunities in farming was likely to be sluggish.

They also had good practical reasons for being cautious. The long-term trend of wheat prices, for example, was not immediately apparent, because prices fluctuated wildly from year to year and, somewhat more predictably, from market to market and month to month. Furthermore, high prices were no encouragement when they were merely a response to partial crop failure. Even when landlords felt sanguine about the prospects of wheat, hiring *maîtres-valets* was not their only possible choice, particularly since that choice also had undesirable side effects. After all, *maîtres-valets* may have been cheaper to hire, but they also had less personal incentive to work effectively than sharecroppers whose family's entire future was tied to a successful wheat crop. *Maîtres-valets*, therefore, required much more continuous and tedious supervision. Provided the labor supply was plentiful, an easier, if crasser, alternative might simply be to alter the terms of sharecropping contracts to increase the landowner's share of the wheat harvest.

What seems to have happened in Loubens is that landowners embraced both alternatives: some shifted from leasing their farms on shares to hiring *maîtres-valets* to work their farms; others reduced their sharecroppers' share of the crop and some may have done both. The record is none too clear. In the eighteenth century, working the land with *maîtres-valets* seems to have been a recent and very tentative innovation that

spread outward from the suburbs of Toulouse, geared to urban proprietors eager to supervise more closely their agricultural operations.[36] In
late eighteenth-century Loubens, for instance, we can document the
presence of only three *maîtres-valets* at any one time, nor did wheat profits necessarily account for their presence.[37] In the case of the Gounon-
Loubens estate, the only farm worked by *maîtres-valets* was La Borio,
considered the estate's home farm and the only one accredited as noble
land, and therefore exempted from most land taxes.[38] Putting *maîtres-
valets* on this farm may merely have reflected a wish for closer supervision over this largest and most productive of the Gounon farms than leasing on shares afforded; it may also have made economic sense, because
this tax exemption made the use of sharecroppers, who traditionally paid
half the taxes, less advantageous. On the La Borio farm there were no
taxes to share and hiring *maîtres-valets* was undoubtedly cheaper.

Just how extensive *maître-valetage* became in nineteenth-century
Loubens, long after tax privileges had been rescinded, we cannot say.
The surviving accounts of the Gounon-Loubens and of the Hourlier estates provide contradictory examples. In the case of the Gounon holdings, as late as 1819, the balance between sharecroppers and *maîtres-
valets* was exactly what it had been in the 1770s: the tenants of La Borio,
the home farm, were *maîtres-valets*; the other seven farms were in the
hands of sharecroppers. By 1844, the balance had shifted: six of the eight
farms were now being worked by *maîtres-valets*. Then, by the 1860s, the
pendulum swung back: La Borio, the traditional home farm, was the first
to revert to sharecroppers and over the next thirty-five years, every one
of the eight farms of the Château of Loubens replaced its *maître-valet*
with a sharecropper.[39] By contrast, on the five of the eight farms of the
Hourlier estate for which nineteenth-century accounts have survived,
sharecroppers were never displaced at all. Though some shift in the tenancy system did occur, evidently the temporary triumph of *maître-valetage* was never as complete at Loubens as elsewhere in the Lauragais.[40]

In the long run, the rise and decline of *maître-valetage* during the
nineteenth century may indeed have been a landlord response to trends
in wheat prices, but it was a response made possible by an excessive
supply of labor that gave landowners the upper hand. As long as wheat
seemed extraordinarily profitable, landlords had every interest in shifting
to a system, like *maître-valetage*, that increased the amount of wheat that
they could sell. When long-term wheat prices turned sour, as they did
after 1870, coincidentally just as emigration was liquidating the rural labor surplus, landlords went back to a system that shared the risks, as
much as the profits, of agriculture.[41]

The return to sharecropping was made easier for Loubens landlords
because the status of sharecroppers had in any case deteriorated since

the end of the eighteenth century. In the first place, nineteenth-century sharecropping agreements put the landlord in the driver's seat by undercutting the traditional security tenants had enjoyed. In contrast to the previous century, when leases had been drawn up for two, four, and even six years, renewable one-year agreements now became the rule.

Second, when in the 1790s the French revolutionaries had abolished the hated tithe owed to the church, they handed it over to the landowners and not to their tenants. From the sharecroppers' vantage point, the tithe never disappeared: all that changed was the identity of the recipient: landlord rather than parish priest and cathedral chapter. Consequently, even though, as before the Revolution, the sharecropper at first retained the same 45 percent of his net production as his share, by "helping himself" to the tithe, his master had increased his part from 45 percent to 55 percent. On the Gounon-Loubens farms this arrangement lasted into the twentieth century and might have lasted longer had it not been for a growing labor shortage. The tithe on oats was reluctantly abandoned in the mid-1890s, but it took the farm labor crisis of World War I, and the disastrous harvest of 1915 in particular, to stop the collection of the "tenth" on wheat once and for all.[42]

Third, during the course of the nineteenth century, the traditional division of farm income shifted in the owner's favor. Between the 1830s and the 1870s there is evidence of some owners appropriating two-thirds (instead of the customary one-half) of the profits on sheep and livestock or, in other instances, two-thirds (actually, counting the tithe which they also collected, 70 percent) of the wheat crop.[43] Evidently, as rural population pressure built up in the first half of the century, landlords could take advantage of the land shortage to exact harsher terms from would-be tenants. Until the 1870s landowners also had the positive incentive of rising grain prices to spur them on.

The simultaneous decline of wheat prices and the emptying of the countryside changed bargaining terms in the late nineteenth and early twentieth centuries. The tithe was gradually abandoned by landlords and the division of the crop became what it had never really been in practice before, namely fifty-fifty after seed had been set aside and their share of the grain doled out to the harvesters. What is more, sharecroppers once again superseded *maîtres-valets*. By the first decade of the twentieth century, Jerôme Olognon, the conscientious steward of the Gounon-Loubens estate, became quite agitated when a "good" family of sharecroppers gave notice, although he remained confident that, with patience, he could find adequate replacements. By the 1920s, that confidence was eroding: on occasion, Olognon had to put a good face on taking on untried tenants without a farm implement to their name. It was only such marginal farm workers, requiring a landlord's help until their first crop was gathered, who became *maîtres-valets* in the decade after World War I.[44] By the 1930s, while most peasants continued to be sharecroppers at Loubens, not a single *maître-valet* was left in the *commune*.[45] While no one had been paying attention to such changes, the institution had quietly faded away.

Because the first Gounon-Loubens was a businessman who wanted to know exactly what he was buying, our information as to how agriculture was conducted in late-eighteenth century Loubens is much more precise than it was in any other of the four villages that we have examined. Just as Gounon insisted on learning exactly what the lordship of Loubens was worth in the way of manorial dues, he also made sure he was told how much grain his new farms could be expected to produce and what the customary crop rotation was. He naturally also had a complete inventory of livestock drawn up, but only bothered to list the farm equipment for the home farm, La Borio, a list that should have been complete, since, as we mentioned, the farm was worked by *maîtres-valets*, servants who could not be expected to furnish equipment of their own.

In the Loubens of 1771, custom called for a three-year crop rotation, which meant that the plow land of any farm was divided into three parts, known as *soles*.[46] Each of these three *soles* underwent a parallel crop rotation over a three-year cycle, but they were deliberately out of synchronization, so that in any one particular year, each *sole* was being put to a different use. The exception to this system of shifting crops from year to year was land in natural meadows and in vineyards. By contrast, other fields went through three-year cycles beginning with wheat, followed the second year by variable proportions of corn, horsebeans (*fèves*), regular beans, oats, barley, or other grains. The combined market value of the second year's crops was generally reckoned at one-half of the first year's wheat crop. For the third year, the rotation called for fallow, which meant that at any time one-third of the plow land lay unused. Just like cropland, fallow was plowed four times, twice in length, twice in width—the customary *quatre façons* often specified in eighteenth-century sharecropping leases—but nothing was sown on fallow for that year. This fallow served the double function of getting rid of persistent weeds (in an age before herbicides) and of giving the land a rest in preparation for sowing wheat, the master crop.

This crop rotation made few allowances for those fodder crops on which grass-eating animals depended. Save for small quantities of horsebeans and barley used for fattening cattle, as well as oats fed as a high-energy supplement to work animals, the bulk of the cattle fodder, green or dried, was limited to the grass that grew naturally in meadows and pastures that comprised, to cite a specific example, just 8.5 percent of the acreage of the Gounon-Loubens farms. Given the summer climate of the French southwest, natural meadows could neither be created nor enlarged: they required lowlands in the proximity of a water course. In eighteenth-century Loubens, as elsewhere in the region, a fixed acreage of meadowland therefore put a ceiling on the number of cattle that could be kept on a farm. The safety valve of collectively owned waste lands, common throughout Europe, did not exist at Loubens, nor were there, as at Juzet d'Izaut, communal mountain pastures. The limited number of cattle that could be fed put a ceiling on the amount of manure available for the various grain crops, a shortage which, in turn, accounted for eighteenth-century crop yields that probably were no great improvement over what they had been five hundred years earlier.[47]

This inadequate agricultural productivity was further lowered by farm equipment that was both scanty and crude, not only by modern standards, but by standards of widely known eighteenth-century farming theory and practice.[48] The extent of Loubens's underequipment showed up in the discrepancy between the value of livestock and of equipment on an eighteenth-century farm. Take, for example, the Gounon home farm, La Borio, with about 48 hectares, worked by four yoke of oxen. In 1771, the farm's total livestock was assessed at 1,749 livres, while the value of all farm equipment and farm implements was put at 148 livres—8.5 percent of what the cattle were worth. Moreover, two carts accounted for four-fifths of the farm implements' assessed value. For the rest, the inventory listed four plows, which, at three livres a piece, must have been the wheelless, wooden *araires* that merely scratched a five-inch furrow into the heavy clay soil and, lacking a mouldboard, were hardly even able to turn over the earth as they went. The farm inventory also listed four spades, four light hoes, another hoe with three teeth for manure, as well as yokes and strapping for the oxen.[49] What threshing equipment there may have been would have been kept at the château, to which the harvest was brought, but its eighteenth-century inventories, if any, have not survived.

In terms of quantities produced, wheat and corn were the chief crops. Of the two, wheat was king, because in the marketplace, as already noted, it fetched twice the price of corn. When de Gounon bought his Loubens estate in 1771, a detailed estimate of its agricultural productivity projected an average yearly wheat harvest on superior plow land (28 percent of the 112 hectares that he had just acquired) of about 8.4 hectoliters per hectare. Average plow land, (52 percent of the total) produced 5 hectoliters, and poor land (12 percent) only 3.4 hectoliters of wheat per hectare. This meant that for the estate as a whole, the gross yield per hectare was 6 hectoliters or 480 kilograms of wheat harvested. After seed, this left a net average of 4.4 hectoliters or 350 kilograms of wheat.[50] This compares with estimates of average net wheat yields of 525 kilograms at Saint Cézert in 1821 and 550 kilograms at Buzet in 1836, respectively fifty and sixty-five years later and this on soils that were clearly inferior to those of Loubens. Compared to twentieth-century production figures, the eighteenth-century results obviously pale altogether: in the early 1980s, average wheat yields in Loubens hovered around 4,500 to 5,000 kilograms per hectare, twelve to fourteen times their eighteenth-century equivalents. Whatever the reasons, eighteenth-century agriculture at Loubens was notably unproductive.

Wheat and corn were not the only crops grown. Other grains like *paumelle* (a rough variety of barley), oats, and a mixture of wheat and rye sown on inferior land also had some importance. On a much smaller scale, there were beans, horsebeans, green peas, chick peas, and vetch that provided seed used as pigeon feed. The balance between these various crops varied greatly from farm to farm, with some having a narrower range of crops than others. And certainly from the landlord's viewpoint, his income depended on polyculture. For instance, for the year 1 July 1771 to 30 June 1772 (for which the carefully monitored cash income and expenditures ledgers of the Gounon-Loubens estate have survived) of the total receipts 69 percent came from grain sales, 15 percent from the sale of hogs, 10 percent from beans and other legumes, 3 percent from fowl, 2 percent from wine, and 1 percent from silkworm cocoons. Milk and milk products were glaringly absent from the list. Even though the farms carried on a brisk trade in livestock, the five animals that had to be purchased in the course of that year cost one-third more than the cattle— mostly milk-fed calves—sold. The deficit in hay, considerable quantities of which had to be bought on the outside, was equally glaring. This does suggest that while the polyculture of eighteenth-century Loubens relied on cattle as work animals, apparently neither landowners nor peasants could count on livestock to increase their income.[51]

In 1816, a generation and a half later, the subprefect of Villefranche-de-Lauragais in his annual agricultural report noted that conditions were changing in his district. He pointed out that within the last ten years, the sowing of alfalfa and other artificial fodder crops had become popular, a practice accompanied by widespread liming of the land.[52] This unspectacular announcement heralded the very modest beginnings of agricultural modernization in the Lauragais. What the subprefect failed to make clear was that, generally, the new fodder crops (clover, alfalfa, *sainfoin*, vetch) were being sown on what had previously been fallow, that third of the land left to rest in any given year. Not only were these fodder crops a net gain in useful acreage, but because the roots of these leguminous plants fixed nitrogen into the soil—a process not fully understood until the twentieth century—they regenerated the soil much more effectively than fallow ever had, while simultaneously breaking the fodder bottleneck that had limited cattle raising. At the same time, the new fodder crops were as effective as traditional fallow had been in getting rid of weeds that had built up. In turn, the availability of artificial fodder to supplement the hay gathered from natural meadows circumvented a major limitation of traditional farming. By permitting farmers to raise and keep more cattle than before, the new artificial fodder promoted an increase in the manure supply, the shortage of which had kept crop yields so low in the past.

Loubens seems to have been a good example of what the subprefect was talking about. Surviving estate accounts indicate that, as early as 1815, sowing a few acres of clover and alfalfa had become commonplace. Throughout the nineteenth century, artificial meadows continued to increase, as did expenditures for clover, alfalfa, *sainfoin*, and vetch seed and for the lime needed to make them grow. In contrast to a couple of generations earlier, vetch was now more often being grown as cattle fodder than as pigeon feed, its usual usage in the eighteenth century.

By the second half of the nineteenth century, artificial meadows had become so much part of the conventional wisdom in Loubens that special provisions mentioning them were routinely included in sharecropping agreements. A contract of 1868 spelled out the entire crop rotation for a 15-hectare farm: one-third in wheat; one-third in corn, minus enough land to sow 10 liters of beans and 20 liters of potatoes; and for the last third that formerly lay fallow, one hectoliter of horsebeans, one of fodder vetch, and one of seed vetch were to be sown. This crop rotation is likely to have left less than 2 hectares (one-seventh of the plow land) unutilized in a given year. In 1874, on La Maynade, one of the largest farms in Loubens, the new sharecroppers had to promise that on the land devoted to wheat, they would sow "fourteen acres [*arpents*] of artificial fodder and two acres of *sainfoin*." A year later, another contract concerning La Borio was even more restrictive: "On each of the three fields, there is to be at least one hectare of *sainfoin* and more if necessary, considering that no other type of fodder may be grown without the consent of the lessor. . . . The lessee promises to sow neither barley, nor oats, nor vetch, nor potatoes, nor any other plant but wheat, corn and *sainfoin*." Other contracts had more latitude, such as one from 1882 that called upon the contracting *maître-valet* to prepare land where wheat or oats were to be sown, by putting in a fodder crop the year before. "*Sainfoin*, alfalfa or any other artificial fodder is to be put in, as is the practice hereabout."[53]

The fodder crops reduced the amount of land left in fallow, but they did not eliminate fallow entirely, at least not until after World War II. In fact, compared to other sections of the Lauragais, Loubens seems to have

been somewhat backward in this respect. As early as the 1860s most regional landed proprietors were thought to be following an elaborate crop
rotation that left only one-ninth of the land unproductive at any one time;
a handful of even more progressive gentlemen-farmers had apparently
eliminated "dead" (that is, unproductive) fallow altogether.[54]

Loubens' progress in this area was less spectacular. Compared to the
eighteenth-century pattern that kept one-third of the land unproductive
at any one time, by the 1890s this proportion had been reduced to one-
fifth. In the first half of the twentieth century, land left idle was reduced
further to between one-seventh and one-ninth of the total plow land. In
short, it took Loubens' peasants several generations to catch up to what
were considered the mid–nineteenth-century regional norms, yet if the
gains were excruciatingly slow in coming, they were nonetheless significant in the long run.

What ultimately allowed Loubens farmers to catch up with their more
advanced regional neighbors was the increased use of cheap phosphate
fertilizers in the form of slag, a by-product of iron and steel making.
These "superphosphates," first officially acknowledged in the Agricultural
Inquiry of 1892, were being spread on fields before they were sown to
fodder crops. Yet even the 1892 survey claimed that Loubens peasants
only spread an even 10 tons of superphosphates on meadows previously
fertilized with lime.[55] Even if we take such a suspiciously round figure
seriously, 10 tons for some 600 hectares of agricultural land was no more
than token use of such fertilizer. The introduction of artificial fertilizers
was every bit as hesitant and protracted a process as the effort to eliminate unproductive fallow.

These delays in the use of superphosphates at the level of the individual farm points a finger at some of the shortcomings of the sharecropping
system itself. In the period prior to the War of '14, as the French call it
nowadays, phosphate fertilizer was cheap (Fr 7 to 10 per 100 kilogram
bale) and, with Loubens connected to Toulouse by narrow-gauge railway
after 1908, easy to obtain. In retrospect, retired Loubens farmers agree
that spreading such fertilizer on natural and artificial meadows paid for
itself several times over in the additional fodder produced. Yet when one
examines what was actually done at the time, what is striking is the complete lack of consistency in purchasing and, presumably, applying the
new fertilizer. There were wild variations from one farm to the next, between two successive sharecroppers on the same farm and even from one
year to the next when the tenant remained the same. The most that can
be claimed is that, on the average, more fertilizer seems to have been
used in the ten years following the end of World War I than during the
ten years preceding its outbreak. The 1930s, presumably influenced by
the world economic crisis, were generally a period of belt-tightening

among Loubens sharecroppers, of technical regression, rather than a period of progress. The ups and downs of fertilizer consumption during the first half of the twentieth century dispel any facile notion of a self-sustaining agricultural advance. Prior to the 1950s there was no perceptible agricultural "take-off" in Loubens.[56]

There are plausible explanations for this state of affairs. In the first place, even if the landowner or his steward had pushed for agricultural progress, short of the proprietor taking it upon himself to foot the full bill for fertilizer—open treason against the sharecropping system—neither the landowner nor his agent could force a tenant to spend money on anything not specified in the traditional sharecropping contract. Given the increasing labor shortage between the two wars, in any case, these were hardly opportune times for inserting new and burdensome clauses in such contracts. A farm steward's role, in particular, was to see that sharecroppers did not cheat their master, that the annual accounts were promptly settled, that the plowing, planting, harvesting, and threshing got done at the right time, and that livestock and grain were sold at top prices. Stewards, if Olognon and some of his nineteenth-century predecessors are any indication, were continually preoccupied with the variations in the the grain and cattle markets and not at all concerned with the technical evolution of agriculture.[57]

Second, until about World War I, most heads of household among sharecroppers were functional illiterates—not a group likely to strike out enthusiastically in unfamiliar directions. As late as World War I, even the younger generation of sharecroppers was sometimes unable to speak or understand French, its Occitan ethnic identity cutting it off from the wider world.[58] The sort of determined self-education that transformed agriculture in Buzet-sur-Tarn in the 1960s was completely out of reach for Loubens peasants in the first four or five decades of the twentieth century. Changes in agricultural methods therefore came about without leadership, without training, merely by osmosis, by observing what neighbors did who were themselves no better informed.

Third, sharecropping as a special system of tenancy served as a brake on innovation. I have already alluded to its most obvious feature, the lack of continuity in the management of any one farm, which was the reverse side of the sharecroppers' institutionalized insecurity, formalized by one-year tenancy contracts. Perhaps equally important was that most sharecroppers lived hand-to-mouth, very close to bare subsistence, having to borrow from the farm steward, not only for major expenditures like celebrating a baptism (75 pre-1914 francs), a confirmation (Fr 85) or a wedding (Fr 200), but even when the head of household needed Fr 30 for an eye operation in Toulouse or when the family wanted to send Fr 10 to a son doing his military service.[59] Their debt to the farm steward was often in arrears over several years and, having paid up, they could be thrust back into indebtedness by a chance drought or an epidemic of hoof-and-mouth disease.

Sharecroppers also came under pressure to buy new equipment considered increasingly indispensable as the number of adults working a farm continued to decline. The acquisition of mowers, horse-rakes or reapers, and even reaper-binders became a must in the first thirty years of the twentieth century, investments that landlords were unwilling to undertake on their own and which they were even reluctant to make in partnership with their sharecroppers.[60] It is not surprising that sharecroppers should have been hesitant about spending money on chemical fertilizers. True, prior to World War I on the average their share of the cost of a ton of superphosphates came only to about Fr 40. Yet at eight times the price of a ton of lime, the traditional dressing for meadows, Fr 40 was a lot of money for most sharecroppers. It also happened to be one of the few substantial expenditures over which they had full control. Even though they recognized the advantages of superphosphates, their parents had managed with nothing but lime and if money was tight, so could they.

This glacially slow conquest of fallow, begun with the introduction of liming and artificial meadows in the early nineteenth century, and extended through the widening use of phosphate fertilizer in the early twentieth century, was paralleled by improvements in tooling. We are able to measure this latter progress with great precision by comparing the 1771 equipment inventory of La Borio, the Gounon-Loubens' home farm that we examined earlier, with its 1845 inventory. Such a comparison leads to three instant conclusions. First, the list of equipment in 1845 was three times as long as it was in 1771, altering one's original impression of a technology-starved agriculture. Second, implements and equipment now made up a more substantial share of the farm's capital. Between 1771 and 1845, according to the official assessment, the value of the farm's equipment went from 8.5 percent to 18.7 percent of the assessed value of its livestock. Finally, not only was there more of everything, such as three carts instead of two, but significant technical improvements were conspicuous. By the nineteenth century, for example, not only were there carts with iron axles, but the more recent inventory also distinguished between the old wheelless *araires*, the wheeled plow with mouldboard (valued at twice the price of an *araire*), and the wheeled *iron* plow, assessed at almost ten times the worth of an *araire*. The fact La Borio had four such iron plows was indicative of genuine technological change.[61]

Judging from nineteenth-century inventories of the château, by the 1840s threshing was done by oxen dragging heavy wooden or stone rollers over the grain stalks spread out on the threshing floor. Genuine threshing machines, with oxen still providing the motive power (*machines à manège*), can be documented under "expenditures for threshing" in the accounts of the Gounon-Loubens' La Borio farm as early as 1870.[62] As elsewhere in the southwest, such threshing machinery was owned and directed by harvest contractors, who moved their equipment from farm to farm and whose territory probably took in more than one *commune*.[63] Each farm, in turn, had to provide the oxen that kept the machine going as the animals went round and round, yoked to what amounted to a huge screw impelling the machine's gears.

The substitution of a steam engine to do the work previously done by four oxen made faster, more powerful threshing machines possible. The introduction of this innovation was not without conflict. The oxen-driven machines had cost landowners nothing save one-half of the contractor's fee for the operation of the thresher, because the care and feeding of the animals was traditionally incumbent upon the sharecroppers. By contrast, the new machines ran on coal, an "outside" expenditure shared equally between sharecropper and proprietor. Despite several years' resistance by reluctant landlords such as the Gounon heirs, the superiority of the steam-driven threshing machine wore landlord opposition down: on the Gounon *métairies* the steam engine replaced oxen between 1896 and 1901. At about the same time, harvest contractors promoted the use of grain-sorting equipment that selected the largest wheat kernels for seed grain. This routine was a kind of half-way house between tradition— saving randomly harvested grain for seed—and modernity—using only commercial, brand-name seed produced by specialized growers. The only earlier manipulation of seed grain had been the practice of dipping it in copper sulphate to deter fungus, already a widespread local practice by the 1820s.[64]

Little needs to be said about the evolution of Loubens in the period between the two wars and even during the first years of the second postwar era. There was nothing original about the *commune*'s demographic decline. Between 1911 and 1954, Loubens went from a population of 386 to 281, a loss of 27 percent. Compared to the other villages we have studied, the decline was almost identical to that of Cadours and of Juzet d'Izaut, but considerably more severe than that of Saint Cézert or of Buzet-sur-Tarn. The number of inhabitants of the village itself, that is, the *bourg*, declined more markedly than that of the people in the countryside, where the number of farming families had become stabilized, although the average size of these families continued to shrink. As elsewhere, the rural middle class, mini proprietors, village artisans, and farm day laborers were most likely to pull up their roots during this period.

Until the early 1950s the social and agricultural structure of Loubens remained unchanged. The 1929 agricultural survey had counted thirty-four farmers, including four smallholders whose combined holdings added up to only 4 hectares.[65] The one peasant who had leased two farms from the Gounons during the second half of the nineteenth century had since then somehow acceded to the ownership of one of the farms he had previously rented. One family of brick makers that had owned a tiny family farm of 7 or 8 hectares shifted to cattle dealing during the interwar period and managed by a prudent marriage and timely land purchases to begin what was to be their ascent into the ranks of local farmers of substance. The other four owner-operators were small-fry, either truck gardeners or the last of a long line of local *brassiers*, peasants who spaded their miniholding.

These peasant landowners then and later were overshadowed by the twenty-eight sharecroppers listed in 1929, for even immediately after World War II, their number had remained constant. Their essential condition did not change during these decades, although the extension of electricity to the outlying farms of Loubens was the first inroad of modernity into their daily lives. On the negative side, when the narrow-gauge railroad serving Loubens shut down in 1946, the yearly local hog fair of 4 November, where, as late as 1938, some two hundred pigs had been sold and shipped to Toulouse by freight car, also began to fade away. The four hundred year-old tradition died on 4 November 1961: that day a single peasant leading a single pig stood for four hours under the unfurled French flag, waiting for a buyer who never came. The end of the little railway had also made access to the nearest regional market, Caraman, much more difficult. Unable to make it to the regional markets, for a number of years Loubens peasants had little choice but to sell to motorized cattle and hog dealers who came to their farms and who were reputed to offer less than the market price.

Despite the inconvenience that the lack of public transportation entailed, as in Buzet and Cadours during the same time, the continuing rural labor shortage gave sharecroppers a bit more elbow room. During the interwar period, the last dim vestige of the prerevolutionary tithe was fading into oblivion: until the early 1930s at the Hourlier farms every tenth sheaf was still being carried to the barn of the château, whereupon the grain was solemnly and evenly divided, while all of the straw was ritually attributed to *le patron*. Around the turn of the century, the elaborate prohibition against "outside" work by sharecroppers seems still to have been enforced. By the 1930s, when farm work was slack during the winter months, young males in a sharecropping family expected to look for casual employment on the railroad or in construction. Even though the major provisions of tenancy contracts did not change, the proprietor's hold on "his" sharecroppers was evidently loosening.

The prevalence of sharecropping, and the traditional reluctance of bourgeois and noble proprietors alike to invest in the farms they let out on shares, slowed, but did not stop, agricultural progress. By 1929, at Loubens two-thirds of the farms used the efficient double Brabant plows, virtually all of them had reapers and horse-rakes, but there were less than half a dozen of the more sophisticated and expensive reaper-binders and not one single tractor in the community.[66]

Yet this official picture was somewhat misleading, because agricultural contractors from neighboring villages were regularly hired to do some of the plowing. A number of years before the introduction of American hybrid corn changed the picture in the 1950s, local peasants had discovered that corn did better with deep plowing. Such plowing, which only tractors could perform, gave plant roots a chance to spread downward, thereby improving the corn's chances during dry summers.

Important as grain of every sort was for sharecroppers, the bulk of it was directly consumed by the family itself, as well as being fed to its livestock, swine, and fowl. The one-half of the wheat crop left after dividing with the master was traded for bread at the baker's: one sack of wheat of 80 kilograms against 35 kilograms of bread. Corn and barley, the other two major grains, were mostly used as animal feed. Cash was provided by the sale of calves, so that livestock raising became increasingly important in Loubens. By 1932, not counting calves, the number of cattle totaled some 260, of which only a minority were work animals.[67]

This in turn meant that, increasingly, the interests of proprietor and sharecropper diverged. Tenants lived largely in an economy of self-sufficiency, of auto-consumption, supplemented by the cash provided by the calves they sold. Calves were practically their only window on the market economy. Understandably, sharecroppers attached no great importance to grain prices, since, at best, grain sales played an insignificant role in their family budget. It was a different story for the landlord. Although he continued to take patriarchal pride in drinking *his* homegrown wine to go with *his* home-raised pork or chicken, his commitment to this archaic subsistence economy was shallow. After all, the bulk of his farm income was derived, as it had been in the nineteenth century and earlier, from selling his portion of the wheat harvest on the market. He could not help but be most sensitive to the ups, and even more to the downs, of that market. Sharecroppers and proprietors may have shared crops, but they did not share the same universe.

This distinction of outlook is crucial if we are to understand the social revolution, which, in the dozen years between 1950 and 1962, transformed the fundamental relationship between farming and landed property in the *commune* of Loubens. These changes were much more profound than anything we witnessed in any of our other four villages. This revolution does not lend itself to a neat, monocausal explanation, but certainly it cannot be understood unless we realize that from the landlord's (but not necessarily from the sharecropper's) vantage point, traditional tenant farming in the Lauragais was becoming increasingly unprofitable. The crux of the problem was the unfavorable trend of farm prices in general, and wheat prices in particular, compared to the prices of nonagricultural commodities, a trend that had first become catastrophic at the height of the depression in the mid-1930s. The food shortages of the war and the immediate postwar years had obscured the unfavorable long-term trends. As a local noble from a landowning family explained with only slight hyperbole: "two generations ago on one hundred hectares, a landlord family could live with ease (*bourgeoisement*), while three or four tenant families drew a decent living from the same land. Nowadays one family would be lucky if it made any kind of respectable living from those same one hundred hectares." Traditional agricultural estates worked by traditional tenant farmers had become a poor investment.

One obvious response to these constraints would have been to maintain income by intensifying production, by acquiring mechanized equipment, and by embracing the new agricultural methods that broke with traditional rural practice in favor of a "scientific," heavy-investment, high-yield approach to farming. Here the interaction of landlord tradition, economic stringency, and the limitations of the sharecropping system itself conspired to impede such moves.

In contrast to the southeastern sector in the *département* of Aude that still enjoys the reputation of being agriculturally a generation ahead, the western part of the Lauragais where Loubens is located had a long tradition of landlords investing an absolute minimum in their estates. Land, after all, maintained its reputation as the soundest of investments only so long as its owner could sit back and enjoy an assured income. We saw earlier that in the first half of the twentieth century, Loubens sharecroppers, seeking to use labor-saving devices like reapers or hayrakes, could not count on the famous half and half clause in their contract when it came to buying such equipment: generally landlords refused their help in defraying such expenses. There are still older tenant farmers living in Loubens today who claim that they were eager to modernize in the 1950s, but were stymied by the unwillingness of their *patron* to cooperate. Smaller owners of sharecropped farms, in particular, were likely to oppose sinking their hard-earned savings into what they viewed as the rat hole of modernization.

In many cases, proprietors could simply claim inability to pay. In any number of cases, in the best tradition of capitalist mythology, *métairies* really were in the hands of aging widows. The rent that they collected was their retirement income at a time when France's social security system had not yet been extended to the middle classes. Except for the wartime bonanza of having an assured food supply in the midst of shortages, proprietors' real income from their estates had in any case been declining for a number of years. The reemergence of the unfavorable long-term trend was particularly brutal between 1948 and 1950: in 1948 farm products had still enjoyed a clear price advantage over industrial goods pegged to prewar ratios: while the index of agricultural prices stood at 100, that of industrial prices had only reached 71. By 1950 that ratio had been reversed: the index of farm prices now lagged behind that of industrial prices, 100 to 158.[68] If proprietors had any capital for investment, it probably had not been generated by their landed estates and was not going to be invested there. Moreover, until the repatriated French North Africans broke the ice, risking one's land as collateral in order to invest in farm machinery and fertilizer was as unthinkable to most landlords as it was to peasant farmers at the time. The very notion that the Crédit Agricole, the nearest branch of which was in Toulouse, would entertain a request for such a frivolous loan was considered outrageous. Even if landowners had been eager to invest in modernization, many lacked the means.

There were other instances where modernization foundered on the traditional institution of sharecropping itself. According to his former farm steward, when M. Saint Jean-la-Croix, an agronomic engineer by training, married into one of the three great landlord families in Loubens, he urged his sharecroppers to increase their use of fertilizer in order to get better yields. He met with solid passive resistance. Since their *patron* would collect half of the crop by paying half of the expenses, tenants explained to the steward, while they themselves would pay the other half while doing *all* of the additional work that intensive fertilization involved, they failed to see that this was to their interest. If Saint Jean-la-Croix was willing pay the entire bill for the required new fertilizer, they would reconsider. Deadlock ensued.

More than a new definition of equity was at stake here. Sharecroppers, as noted earlier, lived in a subsistence economy with only a narrow opening on the wider market economy. This meant that their personal savings were likely to be meager or nonexistent. Since they were not landowners having collateral to offer, their ability to borrow was even more circumscribed than that of their proprietors. Equipment loans to tenant farmers did not become current practice until the mid-1960s and loans for working capital—to buy commercial seed, fertilizer, fungicides, herbicides, and the like—never did become part of the regional Crédit Agricole's routine. In the real world, the reluctance of sharecroppers to invest more heavily in their agricultural operation was as much a function of poverty as of resentment.

It is tempting to look upon the legislative revolution in 1945–1946 radically redefining tenancy and sharecropping as the primary cause of the explosion that blew traditional society in Loubens out of the water.[69] Indeed, the laws passed on 17 October 1945 and on 13 April 1946, sponsored by the first (socialist) minister of agriculture to come from small-peasant stock, ushered in a radical revamping of the very notion of private property in land by providing farm tenants both with legally secured tenure and arbitrated rents. Yet when one examines the actual operation of these new laws at the grass roots, the appropriate cliché that comes to mind is that of a long fuse being lit. It was to take a while before the accumulated explosive material would actually ignite.

We have already encountered some of the perverse effects of the tenancy statutes in *communes* like Buzet, where the very security that the law now granted to tenants impelled property owners, particularly those with real estate speculation in the back of their mind, to circumvent the law by illegal year-to-year oral agreements. In contemporary Loubens, the tenant farmers' or would-be tenant farmers' complaint was, that to evade the statute, absentee or retired owners preferred to hire agricultural contractors to work their land—even at a minimal profit—shutting out hard-working farmers who were in dire need of additional soil. By hiring a contractor to do all of the plowing, disking, sowing, fertilizing, and other spraying of chemicals, proprietors retained effective control, although the profit left to them was likely to be small. If, on the contrary, they turned their land over to a tenant farmer, landowners' profit was not much better, while they also effectively lost control over their patrimony.

Our concern here is with the operation of the tenancy legislation only insofar as it affected the relations of landlords and sharecroppers, and, ultimately, the structure of landed property and agricultural practice in Loubens. The major provisions of the law that particularly interested most peasants in post–World War II Loubens were those that increased sharecroppers' security of tenure (the minimum duration of leases was initially raised from one year to three years and later to six years), the right to demand that a lease "on shares" be changed to one fixed in money or produce, a ceiling on the landlord's share of the crop fixed to a maximum of one-third (instead of the customary one-half), and, finally, the right of preemption, which meant that if the landlord put the property up for sale, the tenant in place, provided he could match the best outside offer, had first choice in buying the farm on which he worked. The articles that most impressed landlords in a law that they otherwise detested had to do with the provisions for "resumption," a term that referred to the proprietor's right to resume full possession when the lease expired whenever he or his heir decided to work the land on their own.

Among sharecroppers the effects of the new laws were uneven and could hardly have been foreseen. The longer leases were accepted without difficulty, since the shortage of sharecroppers favored incumbent tenants anyway. The transformation of share-cropping to fixed leases turned out to be much more problematical. Among tenants, everyone agreed that, even aside from financial considerations, being a regular tenant farmer was a great improvement over being a sharecropper. More than income was involved. Like it or not, every sharecropper lived his working life in the shadow of *le patron* or of his bailiff; by contrast, a fixed-rent-paying tenant farmer could expect to be master in his own house.

Yet there were financial, legal, and psychological obstacles to overcome. In the first place, the proprietor had to be compensated for his half investment in cattle, fowl, and other farm animals and in any equipment to the cost of which he had contributed. No special provisions were made for bank loans geared to this conversion, nor did the Crédit Agricole have any precedent to follow. Inability to finance the conversion from either loans or savings meant that some sharecroppers were unable to climb up to leasehold tenancy.

Second, it turned out that the small print of the law was considerably less favorable to the sharecropper than its major headings. According to that small print, if the proprietor refused to accept a conversion of share-cropping to fixed rent tenancy, the tenant could indeed take his case to the regional arbitration commission that the legislation had created. However, he could only force this conversion on his landlord by demonstrating to the tribunal's satisfaction that the proprietor had violated the terms of the *existing* sharecropping contract. This meant that there was no way of forcing the hand of a "good," but conservative, landlord.

This is not to say that in all cases the transition turned out to be traumatic. Some landlords readily complied with the spirit, as well as with the letter, of the new law. As early as 1954, for example, M. Naudy, the son of a sharecropper, assumed his present Loubens farm by leasing it at a fixed rent in kind. It had previously been worked on shares. By-and-by, a persistent sharecropper could wear down even a reluctant landowner: it took the Tillets twenty years to persuade their landlady to change their tenancy from shares to fixed rent, yet she finally bowed to the inevitable.

Third, even where the financial path seemed smooth and the legislative mandate clear, psychological roadblocks could still stand in the way. Theoretically, the conversion from sharing half and half to sharing two-thirds for the tenant and one-third for the proprietor should not have been overly burdensome to arrange. In real life, good relations (*la bonne entente*) might be endangered by pressing for one's legal rights. The expected friction might more than outweigh the anticipated material benefits. More than anything else, this sort of inhibition accounts for the fact that in 1982, thirty-seven years after such contracts had theoretically been outlawed, the two remaining sharecroppers in Loubens continued to share fifty-fifty with their respective landlords.

Finally, in practice the right of preemption, the right of tenant farmers to priority in purchasing the farm they worked if its owner put it up for sale, turned out to be somewhat hollow. Of the twenty-eight sharecroppers residing in the *commune* of Loubens at the end of the last war, only one, Tillet senior, with great difficulty was able to avail himself of this avenue to ownership. The explanation for this generalized failure hinges on the movement of regional real estate prices and therefore on questions of timing. The right time to buy was from the late 1940s through the first half of the 1950s: many local properties were on the market then and their prices were extraordinarily low. In fact, land fetched no more than two or three times its annual agricultural revenue. There were plenty of young sharecroppers eager to become farm owners at the time, but none of them had the savings or the access to credit that buying a farm required. They were simply shut out.

The situation changed rapidly from the 1950s on. The first wave of French North African farmers, made unwelcome in Morocco and Tunisia but compensated for their expropriated lands, hit the regional real estate market, driving regional farm prices up. A second, larger wave of French North African migrants expelled from Algeria in 1961–1962 drove real estate prices even higher. With generous credit from government sponsored agencies, these *rapatriés* ended up with most of the properties remaining on the market. True, by the early 1960s a branch of the Crédit Agricole had opened in neighboring Caraman and even tenant farmers were beginning to be treated with greater consideration. By then, however, most of the good vacant farms had been grabbed up and those that were left were overpriced dregs.

If, on the other hand, we are to grasp the range of landlord responses, we must keep in mind that, while the provisions of the tenancy laws were likely to color their decisions, for them that was not the crux of the matter. Their chief complaint was the declining profitability of the kind of estate agriculture that had been the practice in the Lauragais for several hundred years. For a variety of reasons, traditional landlords were to prove unable to adapt to the burgeoning high-investment "scientific" agriculture. Yet if they ultimately failed, it was not from lack of trying.

In Loubens a whole range of landlord responses may be distinguished. Although these constituted alternative and often mutually exclusive methods of coping, any one proprietor might go through several of them successively. Obviously, not all landowners had all of the same options open to them: neither a busy professional nor an aging widow were likely to hop on a tractor and run their farm themselves. For such people, "direct operation" was not a realistic alternative. Advancing age might also reinforce the conservatism of many proprietors: unless she was financially pressed an elderly woman might not want to deal with unfamiliar arrangements, even if, on the face of it, they made financial sense. Besides, land ownership was more than a quest for maximum income: it also carried the freight of filial piety and local standing.

Psychologically, the most painless response by landowners was not to respond at all, or to respond only so far as they might be compelled by decisions of the tenant-landlord arbitration commission, the *commission paritaire*, which the postwar laws had established. It *was* still possible to dig in and to cling to the traditional sharecropping system, to pay out as little and to interfere as much as possible, to punctiliously collect one's yearly hog, one's monthly brace of chickens, and one's half portion of the grain harvest just as in the good old days. It *was* possible, and Tillet's landlady was the living proof, to stave off for decades the legitimate demands of one's tenants.[70] Yet in the long run, the price of maintaining familiar ways was likely to be prohibitive: income derived from such traditional sources was bound to erode. Increasingly, low investment guaranteed low net return.

It *was* possible, alternatively, for a landlord to bring sharecropping into the twentieth century, provided that he had the capital to do so and could find and keep tenants willing and able to cooperate. If the two existing "improved" sharecropping farms still operating in Loubens are any indication, proprietors were impelled by motives as divergent as unmitigated greed and the fun of participating vicariously in a smoothly running and up-to-date farm operation. If a landlord was willing to consolidate several of the small, traditional *métairies* into one viable farm; if he was consistently ready to invest in modern, but not necessarily new, equipment and in moderate doses of fertilizer and other chemicals; if he gave up on the more obnoxious "feudal" dues in live fowl and fresh eggs to concentrate instead on getting his half share of the farm income easily documented by checks from the cooperative (the proceeds from wheat, sorghum, corn, sunflowers, rapeseed, calves, milk) sharecropping could indeed survive into the late twentieth century, and, if the landlord skimped on "luxuries" like repairs on the tenants' house, could even be made to pay adequately, if not handsomely.[71]

Most tenants' unwillingness to put up with the dependence of sharecropping and many landlords' reluctance to invest induced most of the smaller property owners to switch from crop-sharing contracts to long-term leases defined by fixed quantities of grain, often supplemented by a yearly tribute in hogs and calves of specified weight. Indeed, if the owners were themselves retired farmers, or as municipal councillors, laid claim to a place in the local order of things, they were under intense social pressure to lease their land to local working farmers. In violation of official guidelines, small parcels of land that had potential as building lots were, as elsewhere, likely to be rented out by verbal agreement on a year-to-year basis.[72] If the new arrangements relieved the landlord of the burden of financial participation and continuous supervision, the net return on his investment was low (in 1982, 2 percent after taxes was the locally accepted figure) and the owner lost effective control over his property to the leaseholder, who could normally look forward to staying on as long as he or his heirs were willing to continue farming the property.

A classic solution for an older landowner without heirs was to sell his or her farm *en rente viagère*, a practice which we encountered elsewhere among childless old farmers who wanted to insure that they would be cared for in their last years. At Loubens, *rente viagère* was a purely financial transaction. The tenant provided a down payment, known as *le bouquet*, as well as a yearly rent in kind for as long as the original owner lived. If the *bouquet* was small, the estate owner gained by collecting much higher yearly payments than those officially sanctioned for a normal farm lease: in Loubens, the Daujean family's yearly payments in cash and kind for their tiny 10-hectare farm bought *en viagère* amounted to roughly Fr 11,500 in 1982. Had they been leaseholders, they could not have been charged more than Fr 7,150. For poor sharecroppers with dubious credit ratings like the Daujeans, buying *en rente viagère* was about the only realistic path to land ownership. It was also an actuarial wager on the buyers' part, since all payments ceased with the seller's death. In this regard, the two families in Loubens, having bought from the same *patronne*, had clearly bet on the wrong horse. At age ninety-two, their landlady was still going strong. As M. Castel, the second victim of the lady's longevity, noted with reluctant admiration: "being the widow of a pharmacist, she really knows how to take care of herself."

Yet *rente viagère* was bound to remain something of a special case. Most traditional estate owners, fed up with low returns, were more likely to act like the widow of Loubens's former notary, who, purely and simply, put her farms up for sale. Some local land sales may have been furthered by fortuitous circumstances: in the postwar era two of the three major land-owning families happened to be going through one of the recurring succession crises when estates are divided up among several legal heirs. In any event, Loubens saw a wave of selling in the early 1950s and another one in the early 1960s that such family circumstances dictated.

Most buyers of farmland were working farmers rather than investors in land, but almost all were operators on a sizable scale, like the Faures from nearby Caraman, a family of progressive, prosperous farmers, agricultural contractors, and cattle dealers, who in 1953 picked up 40 hectares of prime bottom land from the Hourlier estate for what in retrospect they admit was a song. By purchase and rental, the Blaquières, former smallholding peasants, former brick makers, former cattle dealers as well, expanded their already enlarged Loubens farm to some 50 hectares. The most spectacular land transfers in the *commune* centered on a family of French expatriates from Algeria, the Castillons, who, beginning in 1962, successively bought up a vast amount of acreage that had belonged to, among others, the Saint Jean-la-Croix and the Hourlier estates. By the mid-1960s the Castillons loomed as large in the emerging new pattern of Loubens agriculture as the Gounons had in the old.

Yet it would be misleading to assume that Loubens' major estate owners went "gentle into that good night." Before their lands were ultimately sold, or, as in the case of what remained of the Hourlier patrimony, leased out, the heads of each of the three dominant land-owning families tried to farm his land directly by exercising his right of resumption as recognized by law. There were powerful social inhibitions to overcome. In the western Lauragais, it was traditionally considered little short of scandalous for a noble or bourgeois landlord personally to get his hands dirty, or, as the French say, *mettre la main à la pâte* (to get his hands in the dough). Moreover, none of the men concerned had ever been working farmers. Of the three, only Saint Jean-la-Croix, an outsider with major estates in eastern Tarn, had academic agricultural training. In retrospect, Loubens farmers still comment derisively on their awkwardness, their ignorance, their bankers' hours, their overreliance on hired help. If the local consensus may be trusted, none of them was cut out to be a farmer. Saint Jean-la-Croix sold out after less than ten years, having farmed three of his half dozen local farms directly. Both Hourlier and D'Estaings, heir to a major part of the Gounon estates, struggled on well into the 1970s, one to retire physically disabled, the other ailing and reduced to farming the last 7 hectares of what had once been the proud domain of the lord of Loubens.

Symptomatic of this social revolution in progress, the decline in the number of Loubens inhabitants accelerated even more sharply than elsewhere: during just eight years between 1954 and 1962, when the pace of these structural changes was most intense, the township lost another quarter of its population. Over the longer haul, between 1952 and 1982, two-thirds of all farms in Loubens disappeared, while the traditional dichotomy between estate owners and sharecroppers was being replaced by a sharper social and economic cleavage among the working farmers remaining. A new social pyramid was emerging in Loubens.

At the bottom of that pyramid were farmers that can only be described as "survivors." More peasants than farmers, what the five of them had in common was that they occupied tiny farms of between 10 and 16 hectares, all but one of them burdened by yearly payments.[73] Their penury added up to primitive living standards, obsolete equipment, a reluctance or an inability to invest, and children unwilling to take up the succession. If these farmers managed at all, it was by dint of scraping to make ends meet, by growing and raising as much of the family's food as possible, by keeping a few cows (the poor peasant's bank) as a hedge against emergencies, by sheltering an aged parent and collecting his pension, or by holding part-time jobs. M. Naudy was the municipal man-of-all-chores, M. Castel helped with electrical installations as his sideline, and M. Daujean, who could not afford a tractor of his own, paid back his neighbor who plowed, sowed, and fertilized for him by helping with the animals and the haying. All of these survivors had begun their lives as sharecroppers and one of them, M. Péhaut, was a sharecropper still. All of them were more or less embittered, convinced that at best they had stood still while others had prospered. Some blamed their stagnation on a Crédit Agricole that catered only to the haves, others on their *patron* who had stymied their wish to modernize, still others on their own lack of gumption when they were young.

The only bright spot was that, somewhat ironically for an administrative measure intended to further modernization, *remembrement* (land consolidation) had helped these least modern of twentieth-century farmers the most. At Loubens as at Cadours, the successful, modern farmers had not waited for the government to rationalize their farming operations. Some intermediate farmers had benefitted, but land consolidation had been the greatest boon to the survivors, whose daily working lives it had made easier.[74] M. Tillet used to have his fields scattered far and wide. He ended up with only two blocks of land, though one of them was further from his farmstead than he appreciated. M. Daujean's 10 hectares had previously been cut up into twenty dispersed fields, some of them with difficult access, others overgrown with spreading hedges. Since 1976, like Tillet, he was down to two fields. Although Péhaut regretted the good hunting that the razing of hedges had ruined and complained about the damage the contractors' heavy equipment had inflicted on the soil, he too admitted that land consolidation, by bringing all of his fields close to his house, had made his daily tasks much more convenient. Yet these survivors were not going to enjoy these new comforts for very long: within fifteen years, all of them will have retired with no one to take their place.

The social pyramid of agricultural Loubens also had a distinct middle tier. Discounting several *forains*, nonresidents farming land within the township like the Faure brothers from Caraman, there were only four Loubens farmers in this middle group. Their farming operations were on a more generous scale, their lives less pinched than those of the survivors. There was the mayor, M. Verdu, who had recently resumed his 30 some hectares when his tenant farmer had retired without a successor and the mayor's adolescent son had expressed interest in becoming a farmer. There was Tillet, who had 42 or 43 hectares, of which he owned, more or less, 20 hectares. "More or less," because as one of five siblings, all of them entitled to their fair share of the property, Tillet was more the tenant-farmer of the Crédit Agricole whose loan had allowed him to pay off his brothers and sisters, than he was a property owner in his own right. In 1982, his quittance payments (*la soulte*) still had twenty years to run.[75] There was M. Noiret, who occupied the larger of the two remaining sharecropped farms in Loubens, and worked his 46 hectares on half shares. There were the rising Blaquières whom we had already met: their 50 hectares, almost half of them leased, included some of the best former meadows now put to the plow. All these were farmers whose equipment was powerful enough to keep up with the increasingly hard-packed soils for which galloping mechanization was held responsible, who did not have to stint when it came to spraying chemicals on their fields, who were up-to-date on their agronomy, although Verdu, a toolmaker turned aircraft executive and the offspring of a long line of cabinet makers, farm stewards, and smiths, was a newcomer to farming. All but Noiret, whose son had died in an auto accident and whose grandson was too young to decide anything, could count on an assured succession.

That did not mean that all of them were on easy street. Take Marcel Tillet's career. His father was the lone Loubens sharecropper turned regular tenant farmer with enough personal savings to buy the 14-hectare farm on which he worked when his *patronne* put it on the market in 1958. By 1961, Tillet junior convinced his father to take out a loan to buy their first tractor. After his father retired in 1973, young Tillet took out another loan from the Crédit Agricole to pay off his brothers and sisters and to acquire additional equipment. Although he was unlucky in his timing—more favorable interest rates had been in force just prior to his loan and were again to be available two years later—inflation gradually whittled down the payments to the level where they would have been had he been an unencumbered tenant farmer instead of an landowner in hock to the Crédit Agricole. Despite his financial burdens, Tillet managed to purchase another 6 hectares and to build a modern home for his family, leaving the old farmstead to his retired parents. He also made the most of *remembrement* in the mid-1970s by accepting hilly slopes suitable for grazing that no one else wanted, but that permitted him to end up with more acreage than before land consolidation. In time he found another dozen hectares to lease here and there and was still looking for more. In helping out his tractorless neighbor, Daujean, Tillet must have been conscious that the former was near retirement age and that these 10 hectares would become available for rental.

Given the heavy burden he bore and the size of his farm, Tillet had not much choice but to continue, like his father before him, to raise cattle as his main occupation. Short on land, but long on debt, his only feasible alternatives were dairying and stock breeding. Grain farmers could count on grossing Fr 5,000 to 6,000 per hectare, whereas raising calves brought in Fr 7,000 to 8,000 for Tillet, who averaged two cows per hectare of land and usually got good prices for his calves. The way the market for calves functioned, however, was something that baffled him: calves raised in the Lauragais were shipped to Italy for fattening, only to be sent back to France where they ultimately ended up as veal on the shelves of French butcher shops. He had no idea as to why they could not be fattened with Lauragais grain.

If Tillet had so far kept his head above water, it was not by much and then only by endless hard work, modest living standards, skillful farming, and careful accounting. Yet were he to count his working hours, his annual profit amounted to less than a year at the minimum wage. This is why during the harvest season, Tillet had long supplemented the family's income by working for agricultural contractors, doing a little informal contracting on his own and, most recently, helping to bring in the harvest of his powerful neighbor, Castillon. Even though he had never gone beyond primary school, from the outset Tillet had systematically experimented to adapt the new farming methods to his land. Agricultural extension agents, he had long ago discovered, failed to take weather conditions and quirks of soil and miniclimates into sufficient consideration when they proffered their advice. Tillet was convinced that he had learned to strike an optimum balance between a moderate investment in chemicals and an optimum yield in crops. He was confident that he had something to pass on to his fourteen-year-old son, enrolled in an agricultural work-study program, who was doing his internship right at home. Men of his own generation who had gambled on success as he had, Tillet believed, even if they had been able to pull it off, had done so by the narrowest of margins. If his son had to start out in similar circumstances, with local land selling at Fr 20,000 to 30,000 per hectare, the youngster's chances would be nil. Even with what the boy could count on inheriting, he would have to scramble to survive.

No one in Loubens would question that the Castillons constituted the apex of the local agricultural pyramid. The family owned and ran a farm more than three times as large as its nearest rival, but their reputation went far beyond the mere size of their operation. They had arrived in

1962, "with only a suitcase to their name," but willing and able to test just how far the Crédit Agricole could be pushed to extend credit. Having in quick succession bought up the Saint Jean-la-Croix estate and the lands inherited by the younger of the D'Estaings brothers, rounded out by a couple of farms belonging to lesser Toulouse absentees, within a brief time, the Castillons provided an agricultural frame of reference for all Loubens. Everyone in the community agreed that, to the extent that high-investment farming had become locally accepted, Loubens farmers had learned the new methods by watching, sometimes from behind hedges, the Castillons at work. Joseph, the current head of the clan (a son was already primed and in training to take over from the father some time in the mid-1990s), was imitated and held up as a warning, resented and liked, envied and respected, kept at arms' length and minutely scrutinized.

Above all, he provided a topic for endless conversation and commentary. Could anyone believe that in a single season of treating *his* fields, *his* tractor had made fourteen passes—or was it sixteen? Could *he* really spread so many expensive chemicals and still turn a profit? Did I realize that in 1981 *he* had obtained a harvest of almost 11 tons per hectare— yes, 11 tons—from one of his wheat fields? Did I know that *he* farmed over 160 hectares and nowadays practically by himself? Had I visited *his* irrigation lake? Had I seen *his* huge wheat storage bins? *Had* I been shown *his* grain dryer? Loubens had mythologized Castillon into a local marquis of Carrabas of Puss in Boots fame.

Yet what the ruling Castillon exuded was neither power nor self-satisfaction. Like members of the established local elite, he spoke cultivated standard French, yet no one would have mistaken Joseph Castillon for the scion of traditional estate owners. His aura was less one of authority or hereditary complacency, than of a Faustian restlessness that translated into an unceasing eagerness for experience. He had forgotten more about farming, local people insisted, than the official agricultural technicians had ever learned.

Castillon had experimented with everything. Alone in the township, he had built his own irrigation lake, irrigating 40 hectares of his land, but concluding from the experience that watering corn in *terrefort* soil created more problems than it solved. On his own, he had invested in drainage to offset the problems brought on by irrigation, duplicating the experience of the farmers of Buzet. He had switched from corn to sunflowers because they required less water and could be harvested a month earlier, while the soil was still firm enough to be able to support heavy equipment. Some years earlier, he had organized a collective hog-raising operation that he had insisted on disbanding as soon as he became convinced that it paid only the minimum wage. He had, as his neighbors so assiduously noted, experimented with hyperintensive wheat cultivation, but remained unconvinced that this was the road to take over the long haul: during the preceding year he had been comparing the results of doing intensive wheat cultivation on one field and low-investment, low-yield cultivation on another. He took his production, marketing and tax conundrums to the regional Centres d'Etudes Techniques Agricoles (CETA), a cooperative self-help organization that we encountered before in Saint Cézert, of which he was the sole Loubens member, but which currently took in some three hundred of the most progressive farmers in eight *départements*. If a problem came up, for Castillon it was bound to have a solution that in its turn would generate a new set of problems to be solved. If the new lord of Loubens was a working farmer, he was also an entrepreneur to the core.

As of 1983, Loubens' agricultural future appeared predictable. The Castillons and the nucleus of middling farmers who could count on heirs were likely to divide among themselves the small farms left without successors. As the coming generation enlarged its farms, it would probably abandon cattle raising and concentrate on grain farming, which took less effort. Cattle, their numbers already down to half of what they had been a generation earlier, would disappear altogether. In turn, no cattle meant no dung and lack of dunging was likely to accelerate the decline in the humus content of Lauragais soils, already down by 50 percent in the previous twenty years. Larger farms would mean heavier equipment; heavier equipment meant more difficult plowing, requiring still more powerful equipment. That vicious spiral was also likely to accelerate. Erosion had also already assumed a visible shape: even though the *commune* was undulating rather than hilly, deep plowing and runoff from the slopes was changing the silhouette of the land. From the window of their house at the edge of the village, the Noirets could clearly see Caraman to the south. When they had retired there fifteen years earlier, Caraman had been invisible, screened by a line of hills. Every older person had similar stories to tell. At a pace never before observed, the land was being worn down.

6

Historical Perspectives

IN looking back on the agricultural development of our five villages, we might start by returning to a question that goes beyond semantics, a question first raised in the introduction to this book: when did these peasants of the French southwest cease to be peasants and become farmers? Just when and on what grounds does it make sense to speak of agriculture in Haute-Garonne as attaining "modernity"? Does this transition imply that local agriculturists have reached some sort of unprecedented plateau, some new and higher equilibrium?

If we define peasants as agriculturists cut off from the wider world, insulated from any market or money economy, the inhabitants of Buzet-sur-Tarn, Juzet d'Izaut, Saint Cézert, Cadours, and Loubens-Lauragais never were peasants from the time that we took up their story in the eighteenth century. By way of taxes, tithes, manorial dues, rent in kind or in cash, and periodic local and regional markets and fairs they were always linked to a universe beyond their village.

Yet such a definition of "peasant" would be frivolous. What made them peasants rather than farmers was a matter of priorities, not of absolutes, their emphasis on the economic autonomy of the household, their over-riding concern with feeding their families with what they themselves grew and raised. Peasants ceased to be peasants and began acting like farmers as soon as in their own minds the commercial aspect of their economic activity began to overshadow the familial one. At that point it begins to make sense to speak of a modernized agriculture even in the absence of perceptible technological or agronomic progress.

How does this definition fit the evolution that we witnessed in our five rural townships? It should be obvious that from at least the early nineteenth century on, significant agricultural changes were taking place, though not at the same pace everywhere. The conquest of fallow by means of artificial meadows, prepared, first by liming and later by fertilizing with pulverized slag, was drawn out over the ensuing 120 or 130 years. By the outbreak of World War II, this process had been almost completed, its progress varying with the characteristics of the soil, the competence of the individual farmer, and, inversely, with the degree of local conservatism. There is little doubt that overcoming the fallow brought the major gain realized in total agricultural production during this period, although new plows and harvesting equipment may have accounted for the most substantial increase in per capita productivity, allowing a reduced number of agricultural workers to carry on.

Parallel to this conquest of the fallow, but seemingly beginning somewhat later in the nineteenth century, was a long period of slow and not particularly steady technological progress. Some innovations, like the introduction of wooden and iron mouldboard plows that supplemented the regional and very traditional *araire*, may have improved productivity as well, as did the selection of seed grain by the end of the century. Most of the new equipment, however, was designed to save labor, or, more precisely, to adjust to a developing labor shortage, rather than to increase farm output. If Saint Cézert is any indication, as early as the 1860s and 1870s such new equipment permitted yeoman farmers who ran their farms with hired labor to trim the number of workers they employed. Elsewhere, the diffusion of agricultural machines and implements did not cause a decline in agricultural labor, but was a reaction to a rural exodus triggered by new opportunities in surrounding cities or in the commercial vineyards of eastern Languedoc. Only in Saint Cézert do farm laborers seem to have been pushed out. This generalization applies to the introduction of the pioneering threshing machines that had triumphed by the turn of the twentieth century, as it does to the newly introduced mechanical equipment for mowing, haying, and harvesting that came into general use by the period between the two World Wars.

This analysis, and the evidence on which it rests, points to the conclusion that until World War II, the self-sufficient, or nearly self-sufficient, peasant farm still remained a realistic goal. Admittedly, most established peasant households budgeted some "outside" expenditures (over and above the unavoidable real estate taxes) that required them to sell what they had produced over and above their family's needs. In fact, during those years the ideal may have been more attainable than ever before, because the continuing rural exodus included smallholders whose land became available for purchase or lease, allowing remaining land-poor peasants to round out their holdings. As some marginal members of the village community left for the city, the technological innovations so widely adopted during the interwar period tended to preserve, rather than to undermine, traditional peasant values and priorities. By replacing vanishing casual labor with simple and inexpensive machines, peasants were able to fend off disruption of their way of life. Prior to World War II, agricultural progress was conservative in its goals as in its effects.

Such a conclusion may be a bit too pat. Here and there within this predominantly peasant culture incompatible new features were, in fact, evolving. Where substantial "peasant" land ownership generated a sizable agricultural surplus, as it did in nineteenth- and early–twentieth century Saint Cézert, the line between peasant and farmer became blurred. Family economic autonomy remained both an ideal and a near reality among *les gros* of that township, yet no one can doubt the alacrity with which these agriculturists adapted to market opportunities. Even before improvements in equipment and techniques became widely available, the prosperous cultivators of Saint Cézert acted more like farmers than like peasants.

In the case of Buzet, the situation was even more clear-cut. The coming of the railroad divided the working members of the agricultural community between those who made a radical break with the past by gearing *all* their activities to national, and even international, markets and those who merely found the railway a useful adjunct to their traditional farming. On the sunny slopes above the village, the *brassiers*-turned-truck-gardeners producing early vegetables for Paris and even London wholesalers became farmers completely dependent on the money economy— even though they relied on the primitive spade as their basic tool. By contrast, the farm operators of the plain, whether owners or sharecroppers, never lost sight of the agricultural self-sufficiency of their households. Yet we must also remind ourselves that in Buzet truck farming proved no irresistible wave of the future: by the 1940s it was disappearing. The dynamic, up-to-date farmers of present-day Buzet are, to the extent that they are of local origin at all, the descendants of the conservative peasants of the plains, rather than of the market-oriented go-getters from the hills.

More archaic, and eventually also doomed to extinction, were the poverty-stricken mountaineers turned tinkers and scissors grinders who, every spring, streamed out of Juzet d'Izaut to seek their family's livelihood on the road. Their incentive, as we saw, was an acute land shortage that made the realization of the peasant ideal—agricultural self-sufficiency— unrealizable. The migrants of Juzet were no doubt attuned to a wider world, but their adaptation was every bit as ritualistic as that of the most tradition-bound peasant. Yet as rural migrants, they did differ from peasants *or* farmers. As permanent emigration gradually drained the village of its excess population and land became more available, the yearly migrants asked nothing better than to revert to uneventful sedentary lives. By the 1920s, for the first time in at least 150 years, Juzet had become a conventional peasant community where people stayed put.

Nonetheless, each in their own way, the yeoman farmers of Saint Cézert, the truck gardeners of Buzet, and the perennial vagrants of Juzet were departures from what was the predominant peasant norm. Even as machinery and mechanical equipment lightened the traditional agricultural chores, even as alfalfa and *luzerne* reclaimed the ancient fallow, down to the eve of World War II, most rural folk who put their hand to the plow retained peasant expectations. The drastic transformation of peasants into farmers did not get started in earnest until the late 1940s or beyond, when the beginnings of motorization opened a new chapter in the agriculture and the social structure of the French southwest.

There is no good reason to rehearse once again the sequence of agricultural and societal changes that, with local variations, could be encountered in all of the villages studied, though in Juzet d'Izaut, as in most other mountain communities, environmental handicaps aborted effective modernization. We saw that, normally starting with the introduction of the first tractors, within a period of fifteen or twenty years, local agriculture went inexorably from low-investment/low-yield to high-investment/high-yield farming increasingly based on state-of-the-art technology and the latest in agronomic science. The tractor ushered in a panoply of new tooling, new seeds, new fertilizers, insecticides, fungicides, and herbicides, new systems of field consolidation, voluntary or imposed, new or rediscovered crops, and an array of what in the regional context were agronomic innovations crowned by sprinkler irrigation and drainage. On good land, wheat production quadrupled; on newly irrigated land, harvests of 9 or 10 tons of the new hybrid corn per hectare became routine and this on mediocre soil where only the foolhardy might have expected as much as 1.5 tons before the war. Dairymen relied on radically new ways of breeding, feeding, and stabling their cows. The number of cows per hectare tripled; the amount of milk per cow quintupled. Although such results reflected what could be and was being accomplished only by "advanced" farmers, this is what counted: progressive farmers were, quite literally, inheriting the earth. Such an unprecedented transformation should not be belittled and trivialized. Whatever its ultimate consequences, it is an astonishing technical and human achievement.

At least down into the early 1980s "progress" in the French southwest had favored the family farm of between 50 and 100 hectares, a size three to six times that of the regional optimum-sized farm before World War II. What made this expansion of acreage possible was the accelerated attrition of marginal working farmers, whose land became available for rental or sale as they retired or died without heirs. The new middle-sized farmers also benefited from the disarray of traditional large estate owners, most of whom, ill-equipped to cope with unfamiliar conditions, reluctantly presided over the dismemberment of their properties.

Although this partial extinction of bourgeois and aristocratic landowners may be over, the elimination of small, middling, and, at least in Saint Cézert, even substantial farms, once their present operators reach retirement, will go on. In all of the communities we inspected, the lack of sons willing to take over the family farm dominated older farmers' outlook. It is true that when I interviewed them, shaking their heads over continued and rising unemployment, quite a few farmers nearing retirement wondered aloud if some of the young might not change their minds and remain on the farm after all as conditions in the cities continued to deteriorate. Yet in 1982–1983, not one of the head-shakers was able to name a single member of the younger generation who had actually abandoned urban employment—or unemployment, for that matter—to return to country living or who had gone back on his or her resolution to leave agriculture. Continued chronic unemployment in France, which has been disproportionally hard on those seeking their first job, may eventually change some minds, but as of 1983 this had not happened in communities with which I became familiar.

If recent history continues on its present course, as farmers without a succession retire, their land will be divided up among local farmers who are still youthful or who can count on someone to take their place. Everyone agrees that to start out by buying up a farm is financially out of the question for most young would-be farmers and that only those who inherit will be able to carry on. If this trend reaches its term, around the year 2000 France should end up with half a million highly productive and highly capitalized farm operations, with an agricultural labor force perhaps one-tenth of the size of the one that existed prior to World War II. Such a radical societal change over only two generations would be nothing short of revolutionary.

However uncomfortable many of us may feel with the questionable enterprise of projecting the past into the future, with historians turning tea leaf readers so to speak, at least the outline of that past itself, of what has been happening since the 1940s on farms in this part of the French southwest seems straightforward enough. If the trends are unmistakable, it is also clear from our study that the impact of modernization on local communities has varied widely. Modernization has been enthusiastically embraced and fiercely resisted. It has lifted the farmers of Buzet from ignorance and penury and propelled them into a life of technical sophistication, affluence, and a sense of control over their lives. It has left the equally hard-working farmers of Juzet d'Izaut defeated and demoralized. Despite their common denominators, the stories of our five rural communities turned out to be anything but uniform. Should we be satisfied with a Gallic shrug and a comment on the irreducible particularity of French villages?

I would suggest that some tentative generalizations *are* possible. At least there are five areas that seem worth investigating: first, the relationship between agricultural resources and the modernizing response; second, the links, if any, between the historical record of a community and its contemporary adaptation to modernity; third, the role played by what might be called societal or generational considerations; fourth, the catalytic role of the outsider or outsiders in this process; and, finally, the local significance of land consolidation in the way communities have adapted to modern conditions.

The experience of Cadours and of Juzet d'Izaut suggest, first of all, a modest variant of Arnold Toynbee's explanation of civilization in terms of stimulus and response. The hypothesis that both the lushest and the most deprived local (or regional) environments will be *least* receptive to agricultural modernization makes a priori sense. Those who are doing nicely because nature dealt them a good hand don't have much incentive to change. Inversely, those who are poverty stricken because nature gave them the short end of the stick will find it overwhelmingly difficult to overcome their natural handicap. The experience of our five villages suggests that there is something to be said for this commonsense hypothesis. The communities that made the most of their opportunities in recent decades are the three—Saint Cézert, Buzet, and Loubens—that faced neither massive natural obstacles, nor enjoyed extraordinary natural advantages.

Juzet d'Izaut, like other Pyrenean communities, was sufficiently deprived by nature that its peasants never had either the financial resources or the necessary self-confidence to bank on modernization. And given the harshness of the environment, in Juzet the young gave their own resounding vote of no-confidence to the future of local agriculture by leaving *en masse*, or, if they stayed, by finding outside jobs. Nor can this response be discounted as some unique fluke: practically all of the agricultural communities in this mountainous canton face a similar plight. At best, modernization at Juzet eased this mountain township's transition to cattle and horse raising as an occupational sideline, probably the only realistic solution for maintaining any sort of local farm activities. Yet the economic base of the village has irrevocably shifted to salaries earned elsewhere, to income generated by the local wholesale meat processors (who, as we saw, had only propinquity as their connection to local agriculture), to the pensions of returning retirees, and to tourist spending.

The case of Cadours is not quite so clear-cut, but the lush environment provided by the possibilities of garlic cultivation gave a breathing space to a whole generation of very small farmers and certainly slowed their elimination. I do not want to claim garlic as the one and only explanation of Cadours' rather hesitant response to modernization. Nearby communities with the same agricultural advantages, but with a nucleus of younger and more dynamic farmers, took pride in being in the vanguard of progress. For better or for worse, garlic served as a partial and temporary buffer against modernization. It permitted Cadourciens the luxury of years of wrangling over land consolidation; it allowed Cadours farmers the satisfaction of turning down irrigation in a fit of collective pique. Garlic provided a safety net for local farmers who sought "to keep up with progress" (*à suivre l'évolution*), but at their own sweet pace.

Whatever weight we may wish to attribute to the natural environment, to what extent can we really relate the course of modernization in a rural community to its particular social and agricultural history? To pursue our second line of inquiry, how great a role has historical continuity really played? The answer seems to be: "not much of a role." Local history does not seem to offer a reliable guide to the collective behavior of contemporary *communes*. Take Cadours as one example. We saw that of the five villages studied, Cadours stood out as a township in which, over the last two hundred years, small peasant proprietors gradually acquired the bulk, though never the totality, of the land and where most present-day farmers are the genealogical descendants of these smallholders. So far, so good. The difficulty arises in trying to translate historical experience into a specific type of behavior in the face of agricultural modernization. Why should we assume, for example, that the smallholding tradition necessarily foreshadowed the brouhaha over land consolidation or the rejection of collective irrigation in the 1960s and 1970s?

In other cases, the historical continuity thesis appears seductive until closer inspection reveals that it is based on demographic illusions or confusions. It would be tempting to argue, for example, that when the railroads provided the opening, Buzet agriculturists pounced on the opportunity by specializing in truck gardening for distant markets; that therefore, when the availability of irrigation one hundred years later provided a similar economic break, Buzet farmers were historically conditioned to leap at such an opportunity. It *is* true that in both cases economic opportunities were seized, but the linkage turns out to be purely fortuitous. As already noted, not one of the successful, modernizing farmers of present-day Buzet-sur-Tarn is a physical descendant of one of the hillside peas and beans growers: their offspring left long ago for Toulouse. Ironically, some of the resentful returned pensioners who defeated the farmers' slate in the municipal elections of 1983 may well be the children or grandchildren of the vanished truck gardeners. In fact, over half of the members of the irrigation and drainage associations—the hallmark of farm modernization in Buzet—are newcomers to the community; North Italians, French North African expatriates, Aveyronnais. History did repeat itself in a way, but the repetition seems irrelevant: the innovators of the second half of the twentieth century are neither genealogically nor historically rooted to the pioneers of the truck gardening boom of the nineteenth century.

At Loubens the changing of the guard is even more striking: the crumbling of an agrarian system based almost entirely on noble and bourgeois landowners relying on sharecropping and similar forms of tenancy has undergone an almost complete turnover. There are, as we saw, still traces of the historical past that remain, such as the two remaining sharecroppers and the handful of aging ex-sharecroppers turned tenant farmers or proprietors on minifarms, whose land will revert to larger farmers as soon as they retire. Within twenty years, these living memorials to a very particular past will all be gone.

Of the five villages that I examined, Saint Cézert is the only one in which history seems to have conditioned the local farmers' response to agricultural modernization. In the first place, despite some newcomers, most of Saint Cézert's farmers belong to long-established local families whose names go back at least one hundred and fifty years in the land registers and in the nineteenth-century lists of municipal councillors. In contrast to Buzet and Loubens, here there is genuine human continuity. Second, most of these families, as noted earlier, have been attuned to market agriculture for almost as long a period. Their willingness to experiment, as in the case of fodder crops in the nineteenth century and commercial wine production in the twentieth century, had long marked off Saint Cézert from neighboring communities. I don't know that Saint Cézert's tradition of trimming its sails to the prevailing economic winds made the latest shift from grapes to irrigated corn and sunflowers predictable, but it at least allows the historian to nod sagely. Yet this is about all it does. It would seem that when it comes to agricultural modernization, historical determinism yields very little in the way of satisfaction.

A third, and more helpful, approach to the problem of local diversity may be to look for certain preconditions to agricultural modernization that seem to show up again and again: the generational structure of a community, the commitment of the rising generation to agriculture as a way of life, and, more nebulously, the quality of local leadership.

We may start by noting that there are elements of social change that cannot be readily differentiated from one rural community to the next: everywhere agricultural modernization seems to have promoted, or accelerated, cultural change at the family level by raising the bargaining power of children and wives. Yet in a sense, this was also part and parcel of the broader movement of national integration by which national norms, increasingly identified as middle-class values, are diffused among an increasingly homogenized general population. The mechanisms of transmission and inculcation may have changed, the pace of change may have speeded up, yet recent trends have continued and deepened the process of turning "peasants into Frenchmen" that Professor Weber observed for the half century preceding World War I. In any event, no rural area seems to have eluded this reordering of the family and its ground rules.

Nor should we look for clues in the degree of local acceptance or rejection of agricultural modernization as such. Typically the choice was not between modernization and tradition, although twenty-five years ago the older farmer resolutely clinging to archaic ways of doing things had not yet become completely extinct. Yet the prevalence of stubborn old codgers does not offer much of an explanation. What counted was the fact that farmers over forty-five years of age were willing to invest only if they could count on a younger family member to succeed them. Wholehearted adoption of the new methods normally coincided with a farmer's confidence that someone was ready to carry on after him. This boiled down to a son's or son-in-law's or, exceptionally, a daughter's, decision to stay on the land. Each of these decisions was seemingly made by isolated individuals, yet each was conditioned by the local human environment and their effects were collective. They determined the future—or lack thereof—of a given village community.

How the young answered the question, "to farm or not to farm?" in any one township had a good deal to do with its generational structure, although the problem of succession recurs continually as members of the older generation retire one by one rather than by some collective mass resignation. The very existence of a younger generation that might stay or not stay could not be taken for granted in rural societies where unmarried farmers were widespread and single-child families common. Much depended also on varying local traditions when it came to leaving or not leaving for Toulouse, the usual destination of twentieth-century emigrants. This meant that one village might have a critical mass of the young remaining, while another three miles down the road, and sociologically and demographically undistinguishable, might not. Where the young remaining in the community were too few for effective mutual moral support, both morale and investment in agriculture were likely to sag. This has certainly been the case among the sharecroppers and exsharecroppers of Loubens and some of the polyculturists of Cadours. Low local morale did not deter a few enterprising farmers in these communities from forging ahead on their own, but it has impeded any collective efforts. Even this modest assessment cannot be broadened into some grand general rule: for instance, at Saint Cézert, the dampening effect that the absence of a rising generation of new farmers would normally have had, has been offset by the EEC's generous bounties to vintners willing to abandon grape growing.

Even more crucial than the existence of a group of potential young farmers in a given township has been the degree of their personal commitment to agriculture as a career. There was widespread consensus among farmers, mayors, and agricultural technicians that until at least the 1950s regional agriculture was victimized by what might be called negative selection. Traditionally, the brightest son of a peasant family became a grade school teacher, so the tale goes, the second smartest a gendarme or a stone mason, while the least able stayed on the farm. There are probably still some farmers in their fifties or sixties who fit such a stereotype, but the twenty years of urban full employment from the early 1950s to the mid-1970s drastically changed the picture. With city jobs available to anyone for the asking, farming abruptly became a career choice, in fact, often a vocation. However, it was much easier to make such a commitment if there was a local peer group in existence that provided mutual support and comradeship. Where such a nucleus of young farmers existed—most brilliantly at Buzet, but also at Saint Cézert, and even, twenty five years ago and less cohesively, at Cadours—agricultural progress was likely to involve formal or informal cooperation within the rising generation.

Just why such generational groupings of committed farmers emerged in some communities, but not in others is not very clear. The quality or potential of the farms that they could expect to inherit had something to do with it, as did the relationships between fathers and sons: in this respect the frequent generational clashes at Cadours, as compared with the parents' deference to their children's wishes at Buzet, represent a striking contrast.

Yet another important variable in accounting for successful modernization was the emergence of effective leadership among young farmers. We saw that in Buzet the combination of a generational peer group committed to staying on the land and the leadership of a Mallorca made for impressive results. Where no one succeeded in asserting such leadership, as at Cadours, cooperative ventures tended to falter and individual farmers were likely to go their own way. At Saint Cézert leadership was also diffuse, yet the local agricultural study group played a role analogous to that of the Cercle des Jeunes Agriculteurs that revolutionized Buzet. Elsewhere a striking individual example might take the place of conscious leadership and might have some effect even in the absence of a cohesive generational core group: at Loubens everyone agreed that in the diffusion of new agricultural techniques a local *rapatrié*, who remained very much a social outsider, played a key role. Without any collective reinforcement, however, imitation alone led to a much more hit-and-miss adaptation to the new agriculture than was the case at Buzet or Saint Cézert.

This leads us to the broader consideration of the outsider's role in the process of agricultural modernization. In its crudest formulation, to identify modernization with outsiders is obviously circular. For example, to "blame" Juzet d'Izaut's agricultural failure on the absence of such outsiders is absurd: given the township's natural handicaps, no *pied-noir* or Aveyronnais in his right mind would ever settle there. It is also clear that generally the postwar tide of agricultural modernization was on the rise prior to the coming of newcomers such as the Aveyronnais and the North African French expatriates. Furthermore, we do have clear instances— Saint Cézert is a case in point—of communities where the major stages of agricultural transformation took place with no significant outside influences. By the time new settlers bought land there, the new high-investment/high-yield agriculture was already firmly in place, though not necessarily in its definitive form. Evidently there was no simple cause and effect relationship between outsiders and agricultural modernization.

And yet in at least three of the communities studied—Buzet, Cadours, and Loubens—local people give considerable credit to outsiders as agents of change or, if not as agents, as catalysts or accelerators of modernization. The case of Buzet and the role of Mallorca is almost too pat, yet in Loubens the influence of Castillon's experimental approach to farming was almost as great. Even in Cadours, where most of the *migrants* failed rather than succeeded, they opened local farmers' eyes to new possibilities by consolidating their fields and to new expectations in bringing modern amenities to their dwellings. The newcomers' example widened local psychological horizons even where it did nothing else. By shouldering the burdens of pioneering, the North Africans in particular allowed local farmers to watch and to learn both from the outsiders' successes and from their failures.

Local imitation was therefore selective. Castillon's farming techniques were never accepted uncritically at Loubens, but he sold the more enterprising local farmers on his experimental approach to agriculture. The improvements that Barnadien made on the house that he bought when he moved to Cadours were not immediately aped, yet they shortly set a standard of the attainable and, in the long run, of what was to become "normal." On the other hand, no one followed the *pieds-noirs* in investing in the caterpillar tractors they brought to the southwest from North Africa, equipment that turned out to be disastrously maladapted to local farming conditions. Nor was economic success the sole criterion for local imitation: no one in Buzet questioned the profitability of the up-to-date dairy farms created by the Aveyronnais, yet when irrigation offered a genuine alternative to dairying, farmers of local origin promptly sold their cows. The local distaste for dairying was not going to abate simply because outsiders could demonstrate that cows could be made to pay handsomely. The example of Dr. Suchich in Cadours also underlines both the impact of the outsider and the limits of his influence. Without the foreign-born mayor, land consolidation would not have come to Cadours when it did, yet if he was able to lead his Cadourciens to the trough, they mulishly refused to drink. In short, no neat assessment of the outsiders' role is possible. While they undoubtedly undermined the local conventional wisdom, they were followed only if and when their direction dovetailed with preexisting local inclinations.

Studying agricultural modernization at the level of village communities also throws an unexpected light on the process of state-sponsored field exchanges and land consolidation with which we have become familiar under the rubric of *remembrement.* I had originally been led to believe that *remembrement* was tantamount to agricultural modernization and that agricultural modernization was inconceivable without *remembrement.* Without land consolidation, French handbooks on the subject persuaded their readers, modern farming methods simply could not be put into effect, while only the authority of the state was powerful enough to override private local interests long enough to achieve the needed recasting.[1] For someone who has no vested interest in *remembrement,* such preconceptions are not likely to survive exposure to local realities.

In the modernizing process official land consolidation played a decisive role only in Buzet-sur-Tarn, where the current generation of progressive, but no longer so young, farmers all agreed that in the absence of land consolidation, they would have refused to take over the parental farm. In a sense, even at Buzet *remembrement* was more important as a symbolic turning point than as a reality. It was a way for the older generation to make a dramatic gesture that, by acceding to what the young had demanded, signaled that the fathers were about to turn over the reins to their sons. This is not to denigrate the convenience of working larger and less dispersed fields, of freeing fields from encroaching hedges, of more effective drainage, and of having one's land grouped around the farmstead. Even at Cadours, as we saw, once tempers had simmered down, most farmers freely admitted that *remembrement* had made their life easier. The same sort of majority opinion could be found in every one of the rural communities where land consolidation had been carried through.

Yet this is a long way from asserting that *remembrement* was the core experience of agricultural modernization. It was not really crucial for a number of reasons. First, the most progressive farmers in a community, as at Cadours, but also in Loubens, were likely to have consolidated their land on their own, by buying up enclaves, selling inconveniently distant fields, bulldozing hedges that broke up what would otherwise have been large fields, and contracting to have drainage ditches dug. Second, where, as in Buzet, a sizable part of the land was divided among a multitude of absentee miniowners owning one or two small fields apiece, according to existing legal rules land consolidation offered no panacea. If someone owned two separate half-acre fields, *remembrement* could join them into a single one-acre field, but such a field still remained inadequate in size for the optimum use of modern farm equipment. Third, as long as the attrition of small farmers continued, land consolidation was, in a sense, always provisional, never definitive. As small farms left without any succession or else without an heir able or willing to work them came on the market to be bought up, leased, or dismembered by local farmers, some of the very farms that *remembrement* had gathered together once again became dismembered. Fourth, the very fact that such "abandoned" land continued to be put up for sale or lease meant that, even in the absence of official land consolidation, farmers could rent or purchase land to round out their holding and thus to undertake their own private land consolidation in a fairly painless fashion. Finally, *remembrement* was only one of several alternatives to attain optimum conditions for modern farming. The same objective of creating a more convenient field system could be achieved by negotiated settlement—or even by the Compagnie d'aménagement des Côteaux de Gascogne–sponsored *multilateral* exchanges, by tax-exempt *bilateral* field exchanges mediated by a specialized technician from the Chambre d'Agriculture, and, if need be, by informal swapping agreements between neighbors, exchanges that never showed up in the official land register.[2] Whatever its undoubted convenience, *remembrement* was neither a magic wand transfiguring the rustic landscape nor even the indispensable key that would unlock agricultural progress.

Although government-sponsored land consolidation did achieve some of its goals in the townships we surveyed, it also revealed an unfamiliar interplay between local authorities and what the French call *l'Administration* (the bureaucracy). *Remembrement* could be made into an unexpected occasion for local authorities to pull a fast one, to maneuver the representatives of the national government into satisfying local needs all in the name of national policy, and, what was even more delightful, largely at regional and national expense. We saw how at Saint Cézert a knowledgeable and politically astute mayor was able to bend the system so that he and his constituents could get what they really were after, namely improved rural and field access roads for which the *commune* had to pick up only 20 percent of the tab. The mayor of Cadours may have been less adroit politically and a more sincere convert to what land consolidation could achieve in its own right, but he too was greatly pleased that the national government and the *département* of Haute-Garonne were practically donating a first-rate network of rural roads to his beloved township.

Equally intriguing were the unintended side effects of *remembrement* in some of the *communes* affected. Most common was the fact that, by balking at paying the (heavily subsidized) outlays that an adequate ditching system would have demanded, local farmers outsmarted themselves.

In Buzet-sur-Tarn and in Saint Cézert, sad experience later impelled farmers to dig more and deeper drains, but with only half of the subsidies to which they had been entitled under the *remembrement* procedure. At Cadours, they merely continued to complain bitterly as to how little had really been accomplished in the way of drainage, forgetting that they themselves had selected drainage "*à la carte*" and chosen the barest minimum. At Juzet d'Izaut, as we saw, field consolidation had the paradoxical effect of hastening the decline of cropping. To some extent cropping had been an adaptation to the existing patchwork of minifields that lent themselves to nothing else. Once the minifields had been consolidated, pasturing cattle or horses proved more economical and less labor-intensive than growing wheat, corn, or potatoes. In the same township, moreover, the creation of spacious pastures for the first time made part-time stock raising practical, thereby going a long way in bringing local agriculture as a full-time occupation to an end. At Loubens, on the other hand, while *remembrement* did very little for the Castillons or the Faures, large, modern-minded farmers who had already bulldozed their hedges and enlarged their fields, it gave the old-style subsistence farms a new lease on life. Indeed, it may have prolonged the careers of just the sort of farmers whose early departure was projected by the very agronomists who extolled land consolidation. In retrospect, some of these consequences may seem predictable, yet in real life no one had foreseen them.

However diverse local responses may have been to this or that aspect of agricultural modernization, "opting out," that is, resolutely and successfully rejecting the new approaches to farming never became a viable, long-term prospect. Just why was agricultural modernization an offer that the peasants of the French southwest could not refuse, even though many of them must have been fearful of embarking on this voyage to an unknown destination? Why *not* say "no" to that first tractor and everything that followed? When I raised this question in interviews a quarter of a century later, the French phrase, *on est pris dans l'engrenage* (literally, we are caught up in the gears) regularly recurred in my farmers' response. What were those gears?

In the course of these local studies, we have encountered all of the major incentives and constraints that shaped agricultural modernization, yet a brief recapitulation may be useful. By and large, and despite the wisdom of hindsight, most farmers remained surprisingly noncommittal about what had decided them to buy their first tractor. The sudden availability of inexpensive, mostly secondhand, machines suitable to small-scale agriculture gets mentioned hand-in-hand with the refrain, "everyone was buying himself a tractor at the time." None of our farmers mentioned the fact that already the very first postwar French Economic

Plan in 1946 had called for the mechanization and modernization of French agriculture and that subsequent Four-Year Plans continued to place a high priority on agricultural investment.[3] Nor did it occur to local farmers to link the sudden availability of mass-produced small tractors to the Marshall Plan.[4] They certainly never speculated as to whether the shock of the German occupation might have had something to do with jolting their nation out of its prewar complacency, although the agricultural learning experience on the part of French farmers who spent the war years working on German farms occasionally was mentioned. In this respect, the gap between microhistory and macrohistory is every bit as wide as the traditional chasm separating microeconomics from macroeconomics.

Yet however dimly farmers of the French southwest still perceived the big picture of the beginnings of modernization in their region and their country, most of them were nowadays fully conscious that the acquisition of a tractor was the first link in a modernizing chain. In retrospect, most of them had no difficulty in tracing a progression that led, link by link, from their first 20 horsepower gasoline tractor to their current high-yield/high-investment farm operations, often on farms three and four times as extensive as their parents'. Was there any more to all this than the economic rationality introduced by the tractor?

The first consideration was familial and returns us to the problem of succession. The rural exodus, as we saw, was no new phenomenon in the French southwest, but the quickening pace of urbanization and industrialization, not only in Toulouse, but in smaller centers like Saint Gaudens, offered an immediate alternative to the rural young. In the postwar era, traditional peasants could not meet the city's competition when it came to income, amenities, work week: their sons and daughters flew the coop, leaving their parents to carry on in the bitter knowledge that they themselves were the end of the line; equally important, young farmers who could hold out nothing better than a traditional peasant life-style would find no marital partners. Peasants who did not modernize were doomed to disappearance. In many cases, the initial phase of modernization was an attempt by the older generation to bribe the younger generation attracted to modernity into staying on the farm by giving it what it wanted and agitated for.

The other major constraints that forced French farmers "to follow progress" were economic. In the first place, as already noted, the prices of farm products during the inflation that had prevailed at least since the late 1930s, and more particularly since World War II, lagged spectacularly behind wages. Not only did this complicate the task of farmers relying on hired labor, but this discrepancy enhanced the attractiveness of urban employment for rural people generally. In some ways, the growing discrepancy in the rate of price increases between farm and nonfarm products, while not quite as extreme, has had an even more profound effect.

In the French southwest in 1952, a well-run, owner-operated farm of 12 hectares may have yielded *twice* the average yearly wage of a male industrial worker. In 1982, even if it had modernized, that same farm would be doing well netting one-half to two-thirds of the yearly income of *one* industrial worker. Indeed, this is why in 1982 a 12 hectare farm was tantamount to living in poverty. To put it in somewhat different terms: in 1952 8 tons of wheat were the equivalent of the average yearly industrial wage; in 1982 47 tons—nearly six times as much wheat—would be the equivalent of the annual income of an average industrial worker. During the intervening thirty years, in real terms industrial wages had quadrupled. Yet during the same period, the output of a southwestern wheat farmer increased by only two and a half times at best, an increase bought by a considerable rise in production costs.[5] If he kept to old methods and did not expand his acreage, he simply could not continue to make a living. Increasing the size of his farm, while intensifying production by employing the new seeds and new methods, was the least painful of several uncomfortable alternatives.

Even these figures understate the dilemma which was, ironically, exacerbated by the French farmers' integration into the immensely complex (and compulsory) French social security system, which combines allocations for children, medical insurance, and old-age, widows', and orphans' pensions, as well as for various minor benefits. Over time, every effort has been made to equalize the social security benefits of such latecomers as farmers with those segments of the population enrolled from the start. However, in terms of organization and financing, social security for farmers has remained autonomous. Since the system is supposed to be self-financing, this has meant steady increases in premiums, though such premiums vary from *département* to *département*, from year to year, and from one farm to the next, largely in proportion to the land-tax assessment rather than to the actual net income.[6] To cite one example among many: in Juzet d'Izaut, one struggling dairy farmer claimed that his social security payments amounted to approximately Fr 5000 per year, between one-fourth and one-sixth of his yearly income, which amounted to about two-thirds of the prevailing legal minimum wage. Even if the Juzet dairyman may have been exaggerating his burden, it was still true that there was no way that a traditional peasant, carrying on agricultural practices aimed at family self-sufficiency, could generate enough cash income to pay such social security taxes. The traditional peasant economy was simply incompatible with the demands of the contemporary social-welfare state.

This brings home the fact that there was nothing "natural" about French agricultural modernization, if by natural we mean to imply the triumph of blind market forces. Social security premiums were only one of the innumerable constraints, subsidies, quotas, incentive payments, tariffs, and juridical ground rules that have guided French farmers since World War II. When it came to inheritance law, France had amended the equal division mandated by the Napoleonic Code to favor the son or daughter who had stayed to work on the parental farm, yet some form of obligtory division among siblings remained the rule. The state, by way of the Crédit Agricole, provided a modest gift of capital in addition to a sizable, subsidized, low-interest start-up loan to young farmers launching their careers, requiring in return that he or she be able to wave a certificate testifying to his or her agronomic training. Professional advice from agricultural technicians attached to the departmental Chambre d'Agriculture (itself regularly subsidized by the state) was his for the asking and was sometimes even volunteered. A young farmer was encouraged to form a cooperative partnership, a GAEC (Groupement agricole d'économie en commune), with siblings or parents by being granted tax and interest rebates, but with the carrot of a supplementary pension, his father was also nudged into prompt retirement at age sixty-five (more recently at age sixty) and into turning over his farm to his heir. By further tax concessions and outright subsidies, our young farmer was made to see the advantages of the cooperative purchase and utilization of farm equipment by way of a CUMA (Coopérative d'utilisation de matériel agricole). If he combined his forces with at least two of his neighbors to build a joint irrigation pond, the national government and the departmental administration jointly picked up almost half of the tab. He was induced to keep accurate accounts that permitted him to deduct the value-added taxes from any equipment bought on his own, while at a nominal cost the state-subsidized Chambre d'Agriculture provided its set of accountants to help him. If he wanted to expand or modernize, he was encouraged to file a development plan with the European Economic Community to qualify for additional subsidized low-interest loans.

Outside forces influenced his daily round of farm activities as well as his entire agricultural year. The varieties of seed that he planted or sowed, specially adapted to his regional soil and climate, had been developed by state-funded agricultural research establishments. When he came to harvest his crops, the harvester might be a combine purchased by a tax-subsidized CUMA, but in any case his crops were marketed through a tax-subsidized cooperative that paid him prices largely determined by political negotiations at the level of the European Economic Community and secured by European tariffs and quota arrangements. Representatives of the agricultural profession played a decisive role in

determining his tax burden, which tended to be geared to the least efficient farmers within his region. If he happened to be among the minority
of farmers whose farm products did not enjoy a price fixed annually, he
was encouraged to form a cooperative marketing board designed to "rationalize" such prices. In response to dire shortages or mountainous surpluses, the French state or the EEC provided financial incentives to
switch from producing rapeseed to sunflowers or from corn to soybeans,
to start a dairy herd or to slaughter it, to plant specific varieties of grapes
or to pull them out. If he became unhappy with the small size and dispersal of his fields, the state-subsidized Chambre d'Agriculture provided
an arbiter free of charge to arrange for field exchanges between neighbors, while the French Revenue Department waived the usual land-
transfer tax on the transaction. If this did not suffice, the Ministry of Agriculture undertook land consolidation paid from national tax receipts,
with national government and departmental administration joining to defray most of the cost of access roads, hedge removal, and field improvements. And, as noted, if he combined with his fellows to plan a cooperative to undertake irrigation or drainage or both, the EEC, the national
government, and the departmental administration jostled each other in
order to underwrite up to 80 percent of the cost. Administrative subsidies may well have become the French farmers' most consistently profitable crop.

Yet this state of affairs invites some skeptical speculation about the future. As long as French agriculture could readily be described as an un-
derutilized resource and French farmers constituted a key portion of the
electorate, departmental, national, and European programs of farm support sold themselves. Now that the argument that "French agriculture
must be fully funded if it is to catch up with more advanced farm producers elsewhere," sounds increasingly hollow, a less bountiful future may
be looming. The planners of 1946 who had counted on France becoming
a major exporter of grain may, if they are still around, wonder as to just
what they have wrought: French wheat and corn exports to other countries of the European Economic Community are contingent on tariff walls
insulating European, including French, foodstuffs from world prices;
French and other European grain exports outside the EEC rely on export subsidies funded by predominantly urban European taxpayers. Both
tariffs and subsidies are supremely vulnerable to, respectively, international pressures and European budget cuts.

   The willingness of the European Economic Community to invest most
of its resources in its agricultural policy was controversial from the onset.
It became more so when Britain, a country with a small, but highly efficient, agricultural sector joined Europe and immediately launched a

campaign to reduce EEC agricultural expenditures. In the long run, the more recent inclusion in the Common Market of countries like Spain, Portugal, and Greece that still have large (although diminishing) agricultural populations has only delayed the inevitable: in the more developed European member states, France included, declining numbers of farmers will, in the long run, translate into waning political influence. The end of an era that began shortly after World War II, characterized by ambitious and costly national and international investment in agricultural modernization, is in sight. The paradox of agricultural modernization is that it ends by undermining its own demographic and political base.

To project a drastic reduction in the level of European, and hence French, support for agriculture is a safe enough bet. To speculate as to just what this lessened support may mean to the farmers of Haute-Garonne is much more hazardous. An end of special subsidies to mountain agriculture is likely to accelerate the already rapid decline in stock raising and dairying in a Pyrenean *commune* like Juzet d'Izaut. Cadours, on the other hand, may be quite unaffected, continuing to bump along on the roller coaster of world garlic prices. And if general French living standards continue to creep up despite continuing mass unemployment, so should the prospects for locally-produced *foie gras*.

The major question concerns the future of high-cost/high-yield agriculture generally. Suppose the European Economic Community no longer guaranteed wheat prices at 121 percent and corn prices at 157 percent of world market levels?[7] Suppose neither France nor the EEC continued to subsidize the bountiful surpluses produced by the new intensive agriculture? In present circumstances, no one can fully tell whether and to what extent contemporary subsidized French agriculture is viable in a bleaker economic and political environment. M. Castillon on his vast Loubens farm may have felt the first gust of the shifting winds when he decided to experiment with minimal investment in some of his wheat fields. Would new circumstances demand a vast scaling back in both outlays and production if price supports and tariff barriers were drastically lowered or disappeared altogether? Would, for example, irrigation continue to be a paying proposition in Saint Cézert or Buzet-sur-Tarn if neither corn nor sunflower prices were kept artificially high by EEC fiat? In any event, the extraordinary edifice erected over the last thirty-five years to support agricultural modernization now seems much more shaky than it did at the time that I interviewed farmers in 1982 and 1983. An agriculture adapted to a very specific political environment may have to be reinvented as that environment undergoes a massive transformation.

This in turn raises an ultimate issue that we have already encountered in several of our villages, namely the extent to which contemporary intensive agriculture in the French southwest is compatible with the preservation of the physical environment on which agriculture has depended in the past. Just as no one really knows how French farmers would cope in a world without direct or indirect official handouts, no one knows whether high-intensity agriculture is ultimately self-destructive. Here and there in this study we have noted some of the warning signals along the way: the problems posed by up-and-down plowing that tractors have made possible and that the redrawing of the field system has often made virtually unavoidable. Everyone agreed that in rolling or hilly terrain, erosion is taking place at an unprecedented rate, particularly in *terrefort* country where plowed fields are exposed to winter rains and storms. Heavier and more powerful equipment, required both because of the increasing size of farms and because the weight of tractors and combines has already packed down the soil thus making small tractors obsolete, is rapidly changing the structure of the soil for the worse. The decline of cattle raising, and thereby the decline of natural manuring and of nitro-gen-fixing fodder crops and the end to the elaborate crop rotation evolved by the early twentieth century, has, as in the Lauragais, dramat-ically lowered the humus content of the land. In irrigated areas, it is still an open question as to whether scientific drainage systems have really solved the long-term problem of excessive watering that leads to the as-phyxiation of the soil. At the time of my interviews, farmers had not yet become aware of the contamination of the ground water through chemi-cal fertilizers, pesticides, and fungicides, but, if trends in the United States are any indication, such awareness was only a matter of time. Ca-dours' precious water supply may by now be an outright public hazard.

Even if French farmers learn to adapt to the new political climate that the very success of agricultural modernization has created, their ability to adapt to the natural environment remains an open question. To that extent, and more than ever, agricultural modernization remains problem-atical: the revolution at the grass roots that we have traced has yet to reach its equilibrium. It seems most unlikely that French farmers will revert to being peasants, yet new pressures may well force these same farmers to relearn their trade. *Il faut suivre l'évolution.*

Abbreviations Used in Notes

ADHG   Archives départementales de la Haute-Garonne
BDR   Bureau du remembrement, DDA-Toulouse
BDS   Bureau de statistique, DDA-Toulouse
DDA-Toulouse   Direction départementale de l'Agriculture de la
                        Haute-Garonne et de la région Midi-Pyrenées
Gounon C papers   Gounon family papers in the possession (in 1983) of
                        Charles-Louis d'Orgeix de Thonel, Loubens and
                        Toulouse
Gounon G papers   Gounon family papers in the possession (in 1983) of
                        Guy d'Orgeix de Thonel, Loubens
INSEE   Institut national de la statistique et des études
                        économiques
RGA   Recensement général agricole
SAFER   Société d'aménagement foncier et d'établissement
                        rural
SAU   Surface agricole utile

Notes

Introduction

1. The major scholarly hubbub has been over the timing of the political integration of the French peasantry into the French political system. More specifically, the focus has been on whether this politicization of the peasantry did or did not take place during the Second French Republic. Much of this controversy has swirled around Eugen Weber's *Peasants into Frenchmen: The Modernization of Rural France, 1870–1914* (Stanford, Calif.: Stanford University Press, 1976), a book that has argued for a much later political awakening. See also his "The Second French Republic, Politics and the Peasant," *French Historical Studies* 11 (1980): 521–50 and "Comment la politique vint aux paysans: A Second Look at Peasant Politicization," *American Historical Review* 87 (1982): 357–389. For the opposing arguments, see Maurice Agulhon, *The Republic in the Village: The People of the Var from the French Revolution to the Second Republic* (Cambridge and New York: Cambridge University Press, 1982) and also his *The Republican Experiment, the Second Republic, 1848–1852* (Cambridge and New York: Cambridge University Press, 1982), Ted Margadant, *French Peasants in Revolt: The Insurrection of 1851* (Princeton, N.J.: Princeton University Press, 1980) and John Merriman, *The Agony of the Republic* (New Haven, Conn.: Yale University Press, 1978) and, more explictly, the latter's review of the Weber book in *Journal of Modern History* 50 (1978): 534–36.

2. The major works of collective synthesis are Georges Duby and Armand Wallon, eds., *Histoire de la France rurale* (Paris: Editions du Seuil, 1975, 1976), volumes 2, 3, and 4 of which are relevant to the period covered in this monograph; J.-P. Houssel, ed., *Histoire des paysans du XVIIIe siècle à nos jours* (Roanne: Editions Horvath, [1976]); large chunks of volume 1, part 2 and of voluem 2 of Fernand Braudel and Ernest Labrousse, eds., *Histoire économique et sociale de la France* (Paris: Presses universitaires de France, 1970, 1977) deal with the French countryside during the period under consideration. To this short list should be added three single-author studies: one is a work of social geography widely regarded as a classic, namely, Gaston Roupnel, *Histoire de la campagne française* (1932; reprint Paris: Plon, 1974); the second is that of a rural sociologist that I found very helpful, Paul Houée, *Les étapes du développement rural*, 2 vols. (Paris: Editions Economie et Humanisme, 1972); the third is John Ardagh, *Rural France* (London: Century, 1983), the best-informed journalistic survey of contemporary French agriculture.

3. Mendras is best known in this country for his *The Vanishing Peasant; Innovation and Change in the French Countryside* (Cambridge, Mass.: M.I.T. Press, [1970]) but also see Centre national de recherche scientifique, Groupe de sociologie rurale, *Les collectivités rurales françaises, études comparatives de changement social* sous la direction de M. Jollivet et H. Mendras, (Paris:

A. Colin, 1971–) and his more recent methodological study, Henri Mendras and
Ioan Mihailescu, eds., *Theories and Methods in Rural Community Studies* (Oxford and New York, Oxford University Press, 1982).

Chapter 1
The Corncribs of Buzet

1. Most of the information on contemporary Buzet-sur-Tarn is based on interviews with the following:

Farmers and retired farmers: Aimé Ayral, Maurice Benazet, Charles Calleja,
Jean-Claude Carrié, Louis Dieudé, André Gorsse, Henri Gorsse, Roger Labranque, Marc and Mlle Millet, Raymond Prunet, Albert, Bruno and Marie-Josette
Ramondenc, Jean-Marie and Bernard Servières, René Tegon.

Officials, surveyors, agricultural technicians: Benet, Berthier, Dorsène,
Garrigues, Meiffren.

2. "Monographie de Buzet-sur-Tarn," (1885), Archives départementales de la
Haute-Garonne (hereafter ADHG) Br 4* 330. Also Léon Dutil, *La Haute-Garonne et sa région. Géographie historique*, 2 vols. (Toulouse and Paris: Privat
and Didier, 1929), vol. 2, 325–28.

3. Georges Frêche, "Dénombrement des feux et d'habitants de 2973 communautés de la région toulousaine," *Annales de démographie historique* 6 (1968–
1969): 410–11. One survival of a seemingly urban tradition at Buzet was the village's continuous patronage of primary education. The teacher who composed the
1885 monograph was able to trace a line of municipally appointed schoolmasters
from the mid–seventeenth century on, one of whom, incidentally, included the
as yet uncanonized Saint Vincent de Paul. In other villages without such a municipal tradition that I have studied—Juzet d'Izaut, Saint Cézert, Cadours, and
Loubens—the earliest appointment of a schoolmaster dated either from the time
of the French Revolution or was a response to the Guizot Law of 1833.

4. Census of 1809, ADHG 9 M 21² A.

5. "Minute du rôle de la communauté de Buzet, Vingtième," 1750, ADHG C
1333. My discussion of the social and property structure is entirely based upon a
detailed analysis of this document. The *vingtième* tax roll has certain advantages
over the use of other eighteenth-century documents available, such as the pre-revolutionary land register (cadastre), the main function of which was to allocate
the traditional basic land tax, the *taille*. Since so-called "noble land" (*les terres
nobles*) irrespective of the social status of the actual owner was exempt from the
taille, it does not always show up on prerevolutionary village cadastres. By contrast, at least in Languedoc, the *Vingtième* was an across-the-board tax levied on
everyone, on landed as on personal property, to which all but paupers were subjected. The Buzet 1750 tax roll has the added attraction in that it clearly distinguishes nobles, bourgeois (by introducing the name with a Sr., M., Mme, Mlle,
Dlle, or Dame), and shopkeepers or artisans (by indicating trades other than agriculture). Generally there were also indications when a property was let out on
shares and when it was let for lease at a fixed rent. The taxable land in the case
of the Buzet *Vingtième* seems to have been confined to the arable, as it only adds

up to 1,300 hectares (as against a post–World War II "agriculturally useful acreage" [the French abbreviation is SAU for *surface agricole utile*] of about 1,570 hectares for the township). There is no indication one way or the other as to whether there might have been some common lands.

6. For the general regional eighteenth-century context, I have relied throughout on Georges Frêche, *Toulouse et la région Midi-Pyrenées au siècle des lumières, vers 1670–1789* (Toulouse: Editions Cujas, 1974), a really magnificent mine of information. A very detailed social analysis of eighteenth-century Buzet utilized by Frêche but unavailable to me is a University of Toulouse D.E.S. thesis by F. Sabatié, "Buzet-Sur-Tarn (1737–1792). Evolution démographique, économique et sociale." For comments on the Buzet beginnings of *maîtres-valets*, peasant families who were remunerated with a fixed yearly wage normally paid in grain for growing wheat, but who, for other crops and for animal productions had the same arrangements as sharecroppers, see Frêche, *Toulouse*, 248–49 and for their infrequency in eighteenth-century Buzet, see p. 349. This form of tenancy will be discussed in greater detail in chapter 5.

7. For the general pattern of peasant land ownership in the vicinity of Toulouse, see Frêche, *Toulouse*, 160–64.

8. "Minute du rôle du Vingtième de la communauté de Buzet," 1750, ADHG C 1333.

9. Frêche, "Dénombrement," 410–11; Census of the Year VIII [1800], ADHG M 643. Population statistics not otherwise documented are taken from the prefectoral *Annuaire de la Haute-Garonne et de l'Ariège* (Toulouse, various dates).

10. "Matrice cadastrale," Buzet-sur-Tarn, 1838, ADHG 1737/146.

11. ADHG 11 M 9. Although there is no guarantee that the nineteenth-century mayors of Buzet (or their municipal secretaries) filled out these annual reports with accuracy, the nineteenth-century questionnaires submitted to them were much more complete in terms of crop yields, seed requirements, and agricultural prices than their twentieth-century equivalents.

12. Such figures cry out for comparison with contemporary crop yields. Rye is no longer grown in Buzet, but prior to drainage in 1982–1983, the conservative average net yield for wheat fluctuated around 3 tons per hectare, rarely reaching 4 tons. In 1983 progressive farmers were convinced that the ongoing drainage of the fields would permit crops of 5.5. to 6 tons per hectare in the immediate future.

In all fairness, the reader should be warned not to take French agricultural statistics *too* literally, as their reliability is always open to question. Historically, although there are ancien régime precedents as well, the demand for systematic yearly agricultural statistics from individual *communes* in the French southwest seems to go back to the requisitioning of grain supplies for the French revolutionary armies fighting along the Spanish frontier in 1793–1794. As late as the pre–World War I years, the request for yearly production statistics was still geared to discovering the amount of local *surplus* grain available in case of national emergency. This tradition is unlikely to have encouraged conscientious and meticulous reporting, but just how much underreporting there was, we have no way of knowing.

Post–World War II agricultural statistics are deficient for different reasons. The widespread refusal by property owners, particularly in suburban or exurban zones, to lease out their lands on an official long-term basis and their preference for illegal verbal year-to-year rentals may lead farmers to come up with fantastic production figures, fantastic only because they claim to be based on land *exclusive* of the illegally rented acreage. These statistics then may be correct as to the total quantities harvested, but the acreage is often understated, which, in turn, inflates the reported crop yield per hectare. Another category of statistics that is inflated for a very different reason is the number of active farmers registered with the Chambre d'Agriculture. A number of inactive property owners insist on being classified as working farmers in order to enjoy the latter's health and retirement benefits. Particularly in larger townships, this systematic overcounting may sometimes inflate the officially registered farm population by as much as a quarter.

13. This figure is based on early–nineteenth century estate records from Loubens-Lauragais, copies of which have come into my possession. The most detailed and reliable survey of mid–nineteenth century Paris is that undertaken and published by the Chambre de Commerce et d'industrie de Paris, *Statistique de l'industrie à Paris résultant de l'enquête faite par la Chambre de commerce pur les années 1847–1848* (Paris: Guillaumin, 1851).

14. The official population statistics for Buzet-sur-Tarn are the following:

1800	986	1896	1204
1809	1096	1901	1181
1831	1241	1911	1255
1842	1320	1921	912
1851	1316	1936	801
1861	1294	1954	818
1872	1316	1954	818
1881	1354	1962	877
1891	1265	1975	1066

1982 figures of 1,200 are estimates by the local mayor. I analyzed the manuscript census lists, that is, the so-called *listes nominatives*, of 1831, 1896, 1921, 1936, 1962, and 1975 (to be found, respectively, in ADHG 9 M 24/5, M 203, M 243, M 1957, M 1925/1, and M 3455/3) and took the totals for these years from the lists. The figures for 1800 and 1809 are in ADHG M 643 and 9 M 2112 A. All the rest are copied from the official *Annuaire de la Haute-Garonne*. In 1809 of a total population of 1,096, 911 lived in the village; in 1885 of 1,354 inhabitants, 740 lived in Buzet village. The trend toward dispersed habitat has markedly accelerated in the twentieth century.

15. "Monographie de Buzet." Prior to the coming of the railroads, Buzet theoretically did have access to river transport. As described by the local schoolmaster in his report of 1885, the Tarn river *was* navigable by raft and shallow-draft boat, even though not by barge or steamship. The city of Albi was about 40 miles

upstream, Bordeaux 200 miles downstream. However, the fact that the 1831 census lists ten carters (*voituriers*) and not a single boatman (*batelier*) suggests that Buzet placed its reliance on overland transport. The river does seem to have been a source of local proteins, as the census of 1896 listed two professional fishermen.

16. "Dénombrement de 1896, liste nominative," Buzet-sur-Tarn, ADHG M 203. Useful though they are for some purposes, local census lists are also extremely frustrating documents. In the first place, few have survived for the period prior to the end of the nineteenth century. Where they have survived, their occupational description of the agricultural population is usually opaque. For nineteenth-century Buzet, for instance, only the *listes nominatives* of 1831 and 1896 have been preserved. Even though the occupational description of various types of peasants may be matched against their land holdings and tax assessments, if any, even this turns out to be inconclusive, because (1) there is no assurance that a label, such as *cultivateur* is even used consistently within any one particular census; (2) because the meaning of the occupational descriptions seems to change from one census to another and from one *commune* to another even within the same general region; (3) because the descriptive terms themselves may be inherently ambiguous. For example, in 1831, 56 men with 109 dependents (of a total population of 1,241) were described as *brassiers*, people, in other words, who worked with their *bras*, their arms. In the parlance of the eighteenth- and nineteenth-century French southwest, this refers to peasants who spaded their holding because it was too small to justify owning a plow and draft animals *or* to propertyless day laborers who used their arms on other people's land *or* to peasants who combined spading their own plot with hiring themselves out. Similarly, an owner-operator, a *propriétaire-exploitant*, may have been a genuine working peasant, but it may also refer to a landowner who personally supervised the farming of his land without himself ever getting his brow sweaty.

17. Frêche, who claims to have read several hundred of these local monographs, complains that they are virtually useless. Even though I can only claim to have read several dozen, I was more struck by the fact that, despite the rigid outline imposed by the Ministry of Education on the local schoolmasters, the quality of the reports varied greatly. One of my criteria for choosing to study the villages that I did was the availability of conscientious and perceptive local monographs. I found the best of these invaluable as a source of information on local social and agricultural conditions in the last quarter of the nineteenth century. I did not take their historical introductions equally seriously, having neither the time nor the inclination to verify their narratives.

18. "Monographie de Buzet." The wines of Buzet-sur-Tarn, the best of which are nowadays sold as *vins de pays*, should not be confused with *Côtes de Buzet* located about 50 miles east of Bordeaux, whose *Cuvée Napoléon* (because of copyright problems recently renamed *Baron Ardeuil*) really does rate among the finest red wines of the southwest, Bordeaux region excepted. Even though Fronton and nearby Villaudric received an *appelation contrôlée* a few years ago, lovers of French wines will recognize "among the best wines of the Haute-Garonne" to be

something of a backhanded compliment. Nor was Buzet's schoolmaster alone in making grand claims: the monograph on Carbonne, located in the valley of the Garonne river, claimed as much for its township. Neither claim finds much of an echo among present-day vintners in either *commune*, yet according to agricultural technicians to whom I put the question, some of the soil of Buzet-sur-Tarn *is* very similar to that of the *Côtes de Fronton*. In the second half of the nineteenth century, this would have placed Buzet wines several cuts above the *gros rouge* just beginning to be mass-produced in lower Languedoc for a growing national market.

19. "Monographie de Buzet."

20. For informative accounts of the phylloxera epidemic from rather different perspectives, see Charles K. Warner, *The Winegrowers of France and the Government Since 1875* (Westport, Conn.: Greenwood Press, 1975): 1–16, and Leo Loubère, *The Red and the White: A History of Wine in France and Italy in the Nineteenth Century* (Albany: State University of New York Press, 1978): 154–73.

21. "Statistique agricole annuelle. Récoltes des l'année 1913," ADHG M 435.

22. By 1932, it had risen to 180 hectares, by 1954, declined to 160. "Statistique agricole annuelle," 1932, ADHG M 1073; "Recensement général agricole," (hereafter RGA) Buzet-sur-Tarn, 1954–1955, Direction départementale de l'Agriculture, Haute-Garonne and Région Midi-Pyrenées (hereafter DDA-Toulouse), Bureau de statistique (hereafter BDS).

23. During the same period in the nearby Lauragais, peasants followed an intricate crop rotation that called for leaving one-ninth of the land plowed but unsown in any one year, a pattern that would be consistent with the figures for Buzet for 1913. The standard French word for fallow (*jachère*) is unknown among old farmers in the southwest even today, since the practice belonged to the everyday world defined by the language of everyday, Occitan, rather than to the official world defined by the French language. The terms that I have come across are *lo graïs d'estiou* in the Lauragais and *lo tiercé* in the Gascon hills. The nineteenth-century official questionnaires did contain a rubric for *jachère* that was invariably left blank in the Toulousain, even though old-timers provided me with details about the practice as carried on as late as the second quarter of the twentieth century. The 1913 questionnaire requested that any fallow land be included in the total arable land (*terres labourables*, unfortunately another ambiguous word in Buzet where the vegetable terraces on the hillside could only be spaded or hoed, not plowed) but did not have a separate category for land left in fallow in any particular year. It is only by adding up the acreage for the various crops listed (excluding peas and beans, most of which were presumably grown on "nonplowable" land) and comparing it to the total arable acreage that I tentatively conclude that relatively little land was left in fallow by 1913—or else the *commune* was falsifying its report.

24. An ideal family farm in Buzet would have had one pair of work-oxen, one or two pairs of work-cows and a horse for cultivating the vineyard. Even in the best of times, only a tiny minority of farms would have realized this ideal, yet in 1913 there must have been over one hundred farms in Buzet that did own a pair

of oxen. Yet only one farmer out of three, for example, owned a horse; others must have worked their vineyards with a work-cow. And I did hear stories of hard-pressed neighbors with a single cow each, sharing them on alternate days in order to get their plowing done. When it came to plowing, because of their superior strength, oxen were clearly preferred to cows. Work-oxen were more expensive to buy, even though they might by and by be fattened and sold at a profit. They were also more expensive to keep than cows, because cows produced a yearly calf that could be sold. The fact that in 1913 the ratio of oxen to cows was 260 to 330 *is* symptomatic of a genuine prosperity.

25. "Statistique agricole annuelle," 1921, ADHG M 760 and 1932, M 1073.

26. "Matrice cadastrale," Buzet-sur-Tarn, 1914 and 1938, ADHG 1126 W 247/248 and Buzet municipal archives.

27. "Dénombrement de 1936, liste nominative," Buzet-sur-Tarn, ADHG M 1057. I have mentioned earlier the fact that owner-operators (*propriétaires-exploitants*) may have included gentleman-farmers or even absentee landlords for whom agriculture was a source of income and prestige, but hardly an occupation. My discussion of *cultivateurs* is based on a random sample of ten individuals so labeled that were checked against the cadastre of Buzet. My inclusion of possible *maîtres-valets* is informed surmise in that I had been told that the Château of Conques, which owned 115 hectares of prime *terrefort* land in Buzet *had* been worked by *maîtres-valets* at least as late as the 1930s. I think it is more plausible that these were lumped among the *cultivateurs* listed by the census takers than to assume that my local informants were in error. The fact that twenty-five families of *maîtres-valets* had appeared on the census list of 1896 makes it even less likely that not a single *maître-valet* should have been left in Buzet forty years later.

28. "Enquête agricole de 1929, Economie rurale," Buzet-sur-Tarn, ADHG M 1022.

29. "Dénombrement de 1936. Liste nominative," Buzet-sur-Tarn, ADHG M 1057. Incidentally, by the 1940s, among Fronton vintners the term *brassier* had come to mean something like "ne'er-do-well," "bum," or "punk." The person who brought this to my attention, himself the son of Fronton vintners, had no idea that the word had ever had any other meaning, and this despite the fact that he had a technician's expert knowledge of the region's agriculture.

30. Georges Duby and Armand Wallon, eds., *Histoire de la France rurale.* Vol. 4, *La fin de la France paysanne*, edited by Michel Gervais, Marcel Jollivet, and Yves Tavernier (Paris: Editions du Seuil, 1976): 93.

31. A useful summary may be found in Guy Cotton, *La Législation agricole* (Paris: Dalloz, 1975): 315–412, a volume in the series "Manuel Dalloz de Droit usuel."

32. RGA, 1954–1955, Buzet-sur-Tarn, DDA-Toulouse, BDS.

33. RGA, 1970–1971, Buzet-sur-Tarn, DDA-Toulouse, BDS.

34. Duby and Wallon, *Histoire de la France rurale*, vol. 4, 573–78.

35. The classic statement dates back to the last years of czarist Russia. For an

English translation see, A. V. Chaianov, *The Theory of Peasant Economy* (Homewood, Ill.: Irwin, 1966).

36. The most perceptive brief account in any language is Gordon Wright, *Rural Revolution in France: The Peasantry in the Twentieth Century* (Stanford, Calif.: Stanford University Press, 1964): 143–82.

37. Cotton, *Législation agricole*, 212–26.

38. For a brief juridical outline, see Cotton, *Législation agricole*, 490–508. Of the several detailed "how to" handbooks I examined, I found Christian Atias and Didier Linotte, *Le remembrement rural* (Paris: Librairies techniques, 1980) the easiest to consult, but others are almost equally satisfactory. For the legal interpretation of land consolidation by the administrative courts, see Maurice Vallery-Radot, *Remembrement rural et jurisprudence du Conseil d'Etat*, 2d ed. (Coutances: Editione OCEP, 1981).

39. My discussion of land consolidation at Buzet-sur-Tarn (as well as in the other four townships discussed in this book) is based not only on interviews with participants and observers, but also on an examination of the voluminous files for each of the *communes* in the archives of the Bureau du remembrement (hereafter BDR) of the DDA-Toulouse.

40. "Enquête sur le projet. Rapport de la Direction départementale de l'Agriculture," Buzet-sur-Tarn, 9 March 1970, DDA-Toulouse, BDR.

41. "RGA," 1954–1955 and 1970–1971, DDA-Toulouse, BDS.

42. See Cotton, *Législation agricole*, 516–25.

43. See Cotton, *Législation agricole*, 230–35.

44. The actual subsidy by the *département* was more complicated in the way it was disbursed: for the first year, the authorities paid the entire amount, thereafter reducing the payments for each successive year until the subsidy was finally reduced to zero at the end of the eighth year.

45. To put this into a regional context, even three years later, of the ten thousand–odd farmers in Haute-Garonne, only sixty individuals, including the twenty-seven Buzetois, had filed EEC-sponsored development plans. As Buzet's Aveyronnais informed me in no uncertain terms, in *their* home *département* eight hundred such plans had been filed!

46. "RGA 1979. Inventaires préliminaires. Fiche communale," Buzet-sur-Tarn, DDA-Toulouse, BDS.

Chapter 2
Requiem for a Mountain Village

1. The discussion of contemporary Juzet d'Izaut is largely based on interviews with the following:

Farmers and retired farmers: Louis Boué, Alain and Simon Cabiro, Robert Daspet, Pierre Dulac, Jean and Pierre Redonnet, Roger Tapié, Carmella and Pierre Tomps.

Technicians, officials, surveyors, nonfarm residents: Barès, Benet, Fillastre, Méda, Servat.

2. "Minute du rôle de la paroisse de Juzet d'Izaut, Vingtième," 1760, 1782–1783, ADHG C 565.

3. "Minute du rôle de la paroisse de Juzet d'Izaut, Vingtième," 1782–1783. ADHG C 565.

4. "Vérification du rôle de la paroisse de Juzet d'Izaut, Vingtième," 1782, ADHG C 565.

5. "Minute du rôle de la paroisse de Juzet d'Izaut, Vingtième," 1761, 1782, ADHG C 565.

6. "Minute du rôle de la paroisse de Juzet d'Izaut, Capitation," 1782–1783 ADHG C 2174.

7. "Minute du rôle de la paroisse de Juzet d'Izaut, Vingtième," 1782–1783, ADHG C 565.

8. "Minute du rôle de la paroisse de Juzet d'Izaut, Vingtième," 1761, 1782–1783, ADHG C 565.

9. No population statistics predating the nineteenth century seem to have survived for Juzet d'Izaut. I have utilized the following figures for this chapter:

1809	716	1891	600
1831	903	1896	560
1836	957	1901	530
1842	902	1926	362
1845	923	1936	277
1851	871	1946	292
1861	731	1954	255
1872	776	1962	188
1881	715	1977	248

The figures for 1845 and 1881 are taken from the "Monographie de Juzet d'Izaut," (1885), ADHG Br 4*. The following are from the ADHG serie M: 1809, 9M 21/3; 1831 and 1836, 9 M 21/9; 1872, M 650; 1896, M 226; 1926, M 261; 1936, M 1051; 1962, M 1925/3. The rest of these figures are copied from the annual prefectoral handbook, the *Annuaire de la Haute-Garonne et de l'Ariège*.

10. "Matrice cadastrale," Juzet d'Izaut, 1828–1913, ADHG 2919/409 and 411.

11. "Enquête sur le travail agricole et industriel," (1848), Canton d'Aspet, Haute-Garonne, Archives nationales C 353/II.

12. "Monographie de Juzet"; "Dénombrement de Juzet d'Izaut," (1872), ADHG M 650.

13. "Monographie de Juzet"; "Dénombrement de Juzet d'Izaut," (1896), ADHG M 226.

14. "Monographie de Juzet."

15. "Monographie de Juzet."

16. "Etat de dénombrement des bestiaux," Juzet d'Izaut, (1809), ADHG HG M 9; "Dénombrement de Juzet d'Izaut," (1872), ADHG M 650; "Monographie de Juzet."

17. The French revolution of 1848 sparked a number of local uprisings in the Pyrenees in which violations of forest rights figured as the major complaint. In one instance the villages of an entire valley sought to secede from France over this issue. See Suzanne Coquerelle, "Les droits collectifs et les troubles agraires dans les Pyrenées (1848)," *Actes du 78e Congrès des sociétés savantes* (Paris: Presses universitaires de France, 1954), 354–64; Jean Pène, "La révolte de Ba-

rousse en 1848," *Revue de Comminges*, 68 (1955): 13–30, 79–91; Louis Clarenc, "Les troubles de Barousse en 1848," *Annales du Midi* 35 (1957): 167–97.

18. "Matrice cadastrale," Juzet d'Izaut, 1828, ADHG 2919/409 and 411. "Monographie de Juzet."

19. "Monographie de Juzet."

20. "Statistique agricole annuelle," Juzet d'Izaut, 1913, ADHG M 435.

21. "Statistique agricole annuelle," Juzet d'Izaut, 1921, ADHG M 762; "Statistique agricole annuelle," Juzet d'Izaut, 1932, ADHG M 1074; "Enquête agricole de 1929," Juzet d'Izaut, ADHG M 1022.

22. "Dénombrement de 1926," Juzet d'Izaut, ADHG M 261; "Dénombrement de 1936," Juzet d'Izaut, ADHG M 1060; "Matrice cadastrale," Juzet d'Izaut, 1936, ADHG 1140 W/395.

23. "Dénombrement de 1926, liste nominative," Juzet d'Izaut, ADHG M 261; "Dénombrement de 1936, liste nominative," Juzet d'Izaut, ADHG M 1061; "Matrice cadastrale," Juzet d'Izaut, 1936, ADHG 1140 W/395.

24. "Statistique agricole annuelle," Juzet d'Izaut, 1921, ADHG M 762, and 1932, ADHG M 1074.

25. Juzet's farmers owned nine tractors and three motortillers at the time. "Récensement général de l'agriculture," 1970–1971, Juzet d'Izaut, DDA-Toulouse, BDS.

26. "Statistique agricole annuelle," 1913, Juzet d'Izaut, ADHG M 435. Summary sheet, Juzet d'Izaut file, DDA-Toulouse, BDR.

27. Official questionnaire filled out by the mayor of Juzet d'Izaut, 11 May 1965, DDA-Toulouse, BDR.

28. Summary statement concerning land consolidation at Juzet d'Izaut by Emile Fillastre, official surveyor and civil engineer, November 1979, DDA-Toulouse, BDR.

29. "Récensement général de l'agriculture," 1970–1971 and "Inventaires préliminaires, RGA, 1979, both Juzet d'Izaut, DDA-Toulouse, BDS.

30. "Dénombrement de 1962, liste nominative," 1962, Juzet d'Izaut, ADHG M 1925/3.

31. This was the number used by the DDA at the time land consolidation was launched in Juzet in 1966, a figure that jibes fairly closely with the sixteen households headed by a farmer and inscribed as such by the census takers of 1962. "Summary Minutes, Commission communale de réorganisation foncière," Juzet d'Izaut, 5 July 1966, DDA-Toulouse, BDR; "Dénombrement de 1962," Juzet d'Izaut, ADHG M 1925/3.

32. "Récensement général de l'agriculture," 1970–1971, DDA-Toulouse, BDS.

33. H. Kreitmann, chief engineer of the Génie rural, to the mayor of Juzet d'Izaut, 14 April 1965, Juzet d'Izaut file, DDA-Toulouse, BDR.

34. Standard *remembrement* progress report cover sheet, Juzet d'Izaut file, DDA-Toulouse, BDR.

35. "Enquête sur le projet," Juzet d'Izaut, report of the DDA-Toulouse, 2 December 1970, DDA-Toulouse, BDR.

36. "Inventaire préliminaire," Juzet d'Izaut, RGA 1979, DDA-Toulouse, BDS.

37. For a useful analysis of Juzet's present and future prospects compared to other communities in the canton of Aspet, see Direction départementale de l'agriculture de la Haute-Garonne, *Plan d'aménagement rural—Canton d'Aspet. Etude des aspects fonciers et agricoles* (Toulouse, August 1982).

38. The last agricultural survey, dating from 1979, listed three farmers with a combined crop acreage of 5 hectares, which has declined since then. "Inventaire préliminaire," Juzet d'Izaut, 1979, RGA, DDA-Toulouse, BDS.

Chapter 3
Changing Course at Sain Cézert

1. My discussion of contemporary Saint Cézert is based on extended interviews with the following:

Farmers and retired farmers: Roger Bouty, André Costes, René Franchini, Henri Gaston, André and Claude Germain, Jean Karlanoff, Henri Labezin, Claude, Jules, and René Labroy, Antoine Moncouet, Robert Roche, Jean Sicard.

Officials and present and former agricultural technicians: Benet, Billières, Delaye, Pradine, Roblin.

2. "Monographie de Saint Cézert," ADHG Br. 7

3. Taking the tax assessment of 1768 as an example, the list of taxpayers numbered 184. As there was no untaxed land in Saint Cézert, the two aristocratic estate owners paid 22.7 percent of the taxes, six substantial land owners 22.5 percent, and ten comfortable landowners (including one lawyer) 17.3 percent, which means that the eighteen top taxpayers (and 7.3 percent of all taxpayers) paid 62.5 percent of all taxes. If anything, this distorts the picture, because the concentration of landed property must have been even greater. Typically, even propertyless households not officially recognized as paupers were taxed, which suggests that the 111 taxpayers assessed at less than two *livres* may have been largely propertyless. Consistent with this supposition, one may surmise that the eighteen top taxpayers may in fact have owned more than three quarters of all land. ADHG E 1535, 1538, 1539.

4. "Liste nominative," 1830 census, Saint Cézert, ADHG M 7.

5. These figures are based on the analysis in the municipal archives of the Saint Cézert cadastre for the years 1830, 1913, 1921, and 1948. I have used the tax rather than the acreage figures on the grounds that the former make allowances for quality as well as quantity and therefore make for more realistic comparisons.

6. "Monographie de Saint Cézert."

7. "Liste nominative," Census of 1906, Saint Cézert, ADHG M 221.

8. The population statistics that I have gathered for Saint Cézert follow. The figures for 1713 and 1790 are taken from Frêche, "Dénombrement." The totals for 1800, 1805, 1809, 1830, 1906, 1921, 1936, and 1975 are taken from the manuscript censuses numbered, respectively, ADHG M 643, M 643, 9 M 21^2 A, M 7, M 221, M 244, M 1057, M 3455/10. The 1861, 1872, and 1881 population figures are taken from the "Monographie de Saint Cézert." The 1968 figure is taken from the minutes of the Commission communale de remembrement. The rest are copied from the *Annuaire de la Haute-Garonne et de l'Ariège*.

1713	54 (hearths)	1901	300
1790	338	1906	300
1800	375	1911	314
1805	337	1921	244
1809	370	1936	262
1830	436	1946	264
1861	418	1954	271
1872	332	1968	221
1881	330	1975	180
1891	314	1977	180

9. "Monographie de Saint Cézert."

10. "Liste nominative," Year XIII, Saint Cézert, ADHG M 643; "Liste nominative," 1830, Saint Cézert. ADHG M 7; "Liste nominative," 1906, Saint Cézert, ADHG M 221; "Monographie de Saint Cézert."

11. "Liste nominative," 1921 census, Saint Cézert, ADHG M 244; and "Liste nominative," 1936 census, Saint Cézert, ADHG 1057.

12. "Récensement général de l'agriculture," 1954, Saint Cézert, DDA-Toulouse, BDS.

13. "Questions sur le produit des récoltes," 1821, Saint Cézert, ADHG M 7; "Monographie de Saint Cézert"; "Statistique annuelle agricole," 1914, ADHG M 300, and 1921, ADHG M 760; Georges Duby and Armand Wallon, eds., *Histoire de la France rurale*. Vol. 3, *Apogée et crise de la civilisation paysanne, 1789–1914*, edited by Maurice Agulhon, Gabriel Désert, and Robert Specklin (Paris: Editions du Seuil, 1975, 1976), 395.

14. "Statistique annuelle agricole," 1914, ADHG M 300; 1921, ADHG M 760; and 1931, ADHG M 1073.

15. "Questions sur le produit des récoltes," Saint Cézert, 1821, ADHG M 7; "Monographie de Saint Cézert."

16. "Monographie de Saint Cézert"; "Questions sur le produit des récoltes," Saint Cézert, 1821, ADHG M 7. As I have been unable to locate regional price series for wheat that go beyond the 1860s, I have relied on a national series charted by Agulhon, Désert, and Specklin in *Apogée et crise*, 396 (vol. 3 of Duby and Wallens eds., *Histoire de la France, rurale*). It is possible that the figures provided by Sapène, the school teacher, presumably referring to the early 1870s, indicated a considerable growth in the grain harvest over the preceding fifty years. The officially provided statistics of 1821 (a year of agricultural plenty) claimed a combined wheat, rye, and rye/wheat mixture harvest in Saint Cézert *one-third* as large as the "good years" reported by Sapène for the period fifty years later. If the official report of 1821 is accepted at face value, we would have to believe that only 208 hectares (of the 652 hectares of arable land inscribed on the land register) were actually in crops during that year. This is particularly hard to accept in the light of Sapène's assertion that the soil of *la rivière* (about one-fifth of the land) was so rich that it never needed to rest. My very hypothetical explanation is that only six years after the end of the Napoleonic Wars (and, presumably, military grain requisitions in the southwest) the mayor of Saint Cézert was chiefly concerned with establishing that no requisitionable surplus was being produced there. I have therefore rejected his crop acreage figures (and the total

for the harvest that can be extrapolated from them) as unworthy of credence, but have accepted his estimates of seed per hectare, ratio of seed to harvest, and per capita consumption of grain, all of which are in line with information that I have come across elsewhere.

17. "Monographie de Saint Cézert." To tell the truth, given Sapène's figures, I am puzzled by the supposed comparative economic advantage of *luzerne* over wheat, even at the all-time low price for wheat, reached in 1901, of Fr 20 per 100 kilograms. Of course, to the extent that *luzerne* was sowed on land that would otherwise have remained in fallow, there was an obvious advantage, but that does not account for the actual decline in wheat production during this period. Sapène's figures for *luzerne* production was twice the tonnage of wheat on the same acreage, yet as hay it fetched only Fr 6 per 100 kilograms. Sapène in his table fails to include income from oats, largely grown for reasons of crop rotation, which, according to an 1883 account book from the La Mothe domain shown to me, fetched about Fr 15 per 100 kilograms, still well below the lowest priced wheat. My conclusion that Saint Cézert farmers produced fodder for *seed* is the only explanation that makes sense, but I cannot document it.

18. "Statistique agricole ordinaire," 1914, ADHG M 300.

19. Both these figures and the quotation are taken from an interesting document in the possession of André Germain, retired farmer at Saint Cézert, who provided me with a copy. The document, dated 3 June 1945, which he was delegated to draw up by fellow members of the muncipal council, was entitled the "Cahier de la Renaissance française" and was requested of all agricultural communities, apparently in preparation of the calling of an "Estates General of Agriculture." Germain recalls taking his task seriously, which makes his figures more credible than those usually forwarded to the prefectoral authorities. The Saint Cézert municipal council duly sent the requested *cahier* to the *chef-lieu de canton*, Grenade, and that was the last anyone ever heard of it. In more recent years, incidentally, agricultural conferences borrowing the same French revolutionary label have actually met in the *département* of Haute-Garonne.

20. "Cahiers de la Renaissance française," Saint Cézert.

21. "Enquête agricole de 1929," Saint Cézert.

22. "Cahiers de la Renaissance française," Saint Cézert.

23. "Statistique agricole *annuelle*," 1914, ADHG M 300; "Statistique agricole annuelle," 1921, ADHG M 760.

24. "Cahiers de la Renaissance française," Saint Cézert.

25. A good starting point for a grasp of the SAFER is the official booklet published by the Fédération nationale des S.A.F.E.R., *S.A.F.E.R., organisation, fonctionnement* (Paris, 1979). A more analytical approach may be found in Cotton, *Législation agricole*, 465–90, with a more critical assessment by the same author in his article, "Quelques réflexions sur les résultats de l'activité des S.A.F.E.R.," *Information agricole*, No. 443, January 1974. My comment on the SAFER playing politics is based upon information developed in interviews.

26. "Extrait du régistre des délibérations du conseil municipal," Saint Cézert, 31 December 1966, Saint Cézert *remembrement* file, DDA-Toulouse, BDR, in which the request is formulated in those terms, though the wording indicates that

the initiative came from an earlier local meeting of the farmers' union. The request was expedited to the prefect of Haute-Garonne, 27 January 1967.

27. "Monographie de Saint Cézert." This statement is made on the strength of perusing several dozen of these local reports—perhaps as many as fifty—when I was in the process of selecting the *communes* that I intended to study. Saint Cézert was the only township in which this field dispersion issue was raised.

28. For the legal arrangements, see Cotton, *Législation agricole*, 508–12. I owe what insight I have into these exchanges to a former *conseiller agricole*, M. Roblin, who in his semiretirement, had become one of several special mediators for *échanges à l'amiable* in Haute-Garonne.

29. Of the four other villages included in this study, from the first meeting of the coummunal commission to the assumption by local farmers of reallocated fields following the completion of the harvest, Buzet-sur-Tarn took three and one-half years, Juzet d'Izaut just short of four years, Cadours five and one-half years, and Loubens-Lauragais three and one-half years. Normally, the related public works—ditching, bulldozing hedges, road construction, and the like—took one or two additional years beyond the date that fields were reassigned.

30. "Cahier de la Renaissance française," Saint Cézert.

31. I had vintners who simultaneously used both hand-pickers and harvesting machines tell me that the costs were comparable and that the difference was only one of convenience and speed, whereas others insisted that machines halved the harvesting costs. The difference of opinion lies, I believe, in different approaches to cost accounting. The winegrower who maintained that costs were about even was probably only counting wages, whereas the one who contended that machines were only half as expensive made a point of mentioning that he had included all aspects of overhead costs, such as social security, meals, maintenance of temporary housing, and the like in his calculations.

32. In the October 1985 issue of an agricultural journal from the French southwest, *Entraid'oc* 18, I chanced on an ad for hand-held pneumatic grapevine clippers which may widen this particular bottleneck by cutting down on the time required for this traditional winter task. Be that as it may, in the winter of 1982–1983, those Saint Cézert farmers tending vineyards were still using the old-fashioned curved *serpette* or, more commonly, hand-clippers.

33. The reason for using these odd starting and stopping dates has to do with a peculiar fickleness on the part of INSEE, the French national statistical institute. I came away from my search impressed with the immense amount and variety of statistical information that INSEE publishes, but depressed by its inconvenience and inaccessibility for anyone taking an historical approach. The only exception to this unhappy rule seems to be instances where INSEE itself has published a historically oriented monograph on a particular topic. Unfortunately agricultural labor is not among these.

In its regular monthly and annual publications, agricultural wages through 1950 were stated in terms of *daily wages*, leaving wide open the question as to how many days per month should be assumed. I arbitrarily figured on 26.5 work days per month. For the period of 1951 through 1973, INSEE reported farm workers' wages in terms of *monthly income*. For the period 1974 through 1977 (the last date I consulted), official reporting was in terms of *hourly earnings*, which I translated into an assumed 180 work hours per month. Each time the system of reporting changed, an otherwise inexplicable chasm marred the data. I found it hard to believe, that, coincidentally with each definitional change of agricultural wage statistics, these salaries should have tumbled from Fr 778 to Fr 408 per month from 1950 to 1951 or shot up from Fr 1,434 to Fr 2,574 from 1973 to 1974. The INSEE data, the only available, are most conveniently found in *The International Labour Office Yearbook*, (Geneva), of which I used the volumes for 1945–1946, 1949–1950, 1957, 1962, 1968, 1974, and 1979. However, for reasons unknown, in subsequent yearbooks, France joins Andorra and San Marino in the short list of countries failing to submit such statistics.

34. The authoritative *Code rural* (Paris: Dalloz, 1981), 302–407, provides a verbatim text of all current legislation which is extremely technical for the lay person. Cotton, *Législation agricole*, 137–145, is clearer on the organization, but offers no clue as to the costs involved. The French social security system is so complex and, despite France's traditions, so curiously decentralized, that it is very difficult to pin down just exactly how much social overhead for their workers Saint Cézert's employing farmers do pay. It apparently varies with the changing balance sheet of the regional (in this case, the Haute-Garonne's) agricultural *caisse* managed under the auspices of the departmental Chambre d'Agriculture.

35. "Recensement général de l'agriculture," Saint Cézert, 1954, 1970–1971, 1979, DDA-Toulouse, BDS.

36. Alain Revel and Michel Gervais, "L'agriculture," in Commission du Bilan, *La France en mai 1981, vol. 2, Les activités productives* (Paris: La documentation française, 1981): 217 for wine prices, 313 for farm workers' incomes. For general cost of living figures (tabulated in terms of purchasing power of the franc according to various base dates), I am relying on an INSEE-based table in "Le pouvoir d'achat du Franc, de 1914 à 1979," *Chambres d'Agriculture* 51 (1980), supplement to no. 662–63.

Chapter 4
Where Garlic Is King

1. Most of the information on agricultural practices in the contemporary period is based on interviews with the following:

Farmers and retired farmers: Mario Bentoglio, Joseph and Jeanne-Marie Calac, Jean and Christian Carbonnel (father and son), Marius Carbonnel (grandfather), Emile and Marie-Louise Cousturian (brother and sister), Bruno Faggion, Enogat Izard, Jean-Louis and Josette Jullian, André Poirier, Henry Raby, André Reulet, Jean Simion.

Officials and technicians: Benet, Georges, Jullian, Peneff, Roblin.

2. The foregoing discussion is based on an analysis of the census "listes nominatives" for 1896, 1921, and 1936, found in the departmental archives under M 203, M 244, and M 1055. There is always a good deal of local controversy as to the number of "genuine" farmers in a *commune*. At Cadours, the discrepancy between the high official figures (supposedly forty-four farms in operation, of which thirty-four have the equivalent of one or more full-time workers) and the

total to which local farmers personally attest is considerable. The latter figure does not reach thirty. For the latest official figures, see the yearly listings issued by and on file at the Chambre de l'Agriculture de la Haute-Garonne, which, in this instance, I copied at the Bureau de statistique of the DDA-Toulouse. In my experience, informal local enumerations, which invariably entail a critical name-by-name assessment, may not be infallible, but they are certainly much more realistic than official listings. Even though most Cadours farms could be run without hiring labor, the fact that as late as 1954–1955, the agricultural census (DDA-Toulouse, BDS) listed thirteen farm hands, suggests to me that in the nineteenth century, the rural population of farm laborers must have been much denser. Together with their families, they may have constituted one-fifth of the local population.

3. As noted in the introduction, this recollection illustrates nicely one limitation of oral history to which I alluded in my introduction to this book: strong impressions may retroactively be "translated" into "appropriate" details that the narrator sincerely believes to be true, yet which are demonstrably false. According to the agricultural census of 1954–1955, two years prior to Barnadien's arrival in the township, there were twenty-two tractors in use at Cadours.

4. The population statistics that I have used for a discussion of Cadours' demographic history follow:

1713	280 hearths	1896	813
1790	192 hearths; 844 inhabitants	1901	802
1800	790	1911	847
1805	790	1921	703
1809	808	1936	632
1831	927	1946	599
1842	1002	1954	590
1852	1035	1962	682
1872	980	1968	737
1881	912	1977	776
1891	847		

The figures for 1713 and 1790 are taken from Frêche, "Dénombrement." The totals for 1800 and 1805 are found in ADHG M 643 and 1809 in 9 M 21₂ A, which are mere summaries, not name-by-name census lists. I have also utilized the following *listes nominatives*: 1896 (M 203), 1921 (M 244), 1936 (M 1055), 1962 (M 1925/1) 1968 (M2747/2). No nineteenth-century census lists have survived for the period prior to 1896 and the census list for 1975 is unaccountably missing. Other population totals are taken from the *Annuaire de la Haute-Garonne*.

5. "Monographie de Cadours," ADHG Br 4° 181.

6. Louis Napoleon's coup d'état had restored universal manhood suffrage, which explains why the number of qualified voters—males twenty-one years or older—is recorded for the *Annuaire* for 1852. The remainder of this paragraph is based on an anlysis of the census lists for 1896, ADHG M 203; 1921, ADHG M 244; and 1936, ADHG M 1055.

7. This notion of a local political elite is derived from conversation with a former mayor who carefully examined nineteenth-century municipal records. There were 299 households in 1885, when M. Faure, the village schoolmaster, drew up

his official "Monographie." According to the *Annuaires*, there were 92 men who met the municipal voting requirements in 1831 and 100 who did in 1842. After 1848, universal manhood suffrage naturally enlarged the number of qualified voters.

8. "Rôle du Vingtième," Cadours, 1758–1759, ADHG C 652. As already indicated in the case of other villages, technically speaking these percentages are based *not* on acreage owned, but on taxes paid, taxes that take quality as much as quantity of land into account. According to a survey conducted during the first year of the French Revolution, there was no privileged land in the township. "Rôle de supplément pour les cî-devant privilégiés," Cadours, 1789, ADHG C 2169.

9. "Rôle de la taille," Cadours, 1790, ADHG C 2170.

10. "Matrice cadastrale," Cadours, 1839–1913, ADHG 1737/153. "Matrice cadastrale," Cadours, 1939–1949, Cadours municipal archives. I also analyzed the "matrices cadastrales" for 1964 and 1981, both of them also kept in the township archives.

11. 1948 is about the last year when land ownership still retained its traditional social significance. The transformation of sharecropping to low-cost, government regulated money leases (although still expressed in set quantities of grain) and the increasing availability of land for leasing as heirs of marginal farmers left for the city, made access to land much more flexible. Hence, as I have noted elsewhere, the advantages of land ownership over tenancy became much less self-evident for working farmers. Before 1950, farmers generally were either tenants or owners. Since the 1950s the majority of farmers combine both and therefore the contemporary distribution of land tells us nothing of either economic or social importance about local patterns of farming.

12. "Rôle du Vingtième," Cadours, 1776, ADHG C 652.

13. "Monographie de Cadours"; Bernard Ufferte, *Notices sur Cadours et sa région—son passé, ses usages, son développement* (Poitiers: Société française d'imprimerie, 1935), 341ff.

14. "Statistique agricole annuelle," Cadours, 1914, ADHG M 299; 1921, ADHG M 760; and 1932, ADHG M 1073. Barnadien's claims notwithstanding, official documentation as well as other Cadourciens' personal recollections suggest that much less than one-third of the land was allowed to rest at any one time. Agricultural practice with respect to fallowing seems to have varied more widely in Cadours than in the other communities that I have studied.

15. The statistics for 1921 registered a total production of only 500 kilograms from 9 hectares planted, a ridiculously low yield and an insignificant total production. In 1932, when total production of garlic was listed at a little over 2 tons, the nearby *commune* of Brignemont already produced ten times as much. I did not come across anyone who claimed to have foreseen the spectacular rise of Cadours garlic (which, incidentally, refers to that produced in the whole canton of that name and not merely in the township itself) since the 1960s.

16. See Ufferte, *Notices*, 306–13, for details on the farmers' union, which was affiliated with corresponding regional and national agricultural organizations.

17. For the distribution of farm equipment in 1929, see "Enquête agricole de 1929. Economie rurale," Cadours, ADHG M 1022.

18. In recent decades the Cadours farmers' union has withered. From the 1950s on, the growth of gigantic farmers' cooperatives that also sold seed and chemicals preempted the union's local service function. During the same period, the creation of a network of agricultural extension agents made the unions redundant as popularizers of new agricultural techniques. Despite its occasional participation in political forays organized at the departmental level against government action or inaction, over the last twenty-five years, the farmers' union at Cadours has been reduced to a largely ceremonial role, in sharp contrast, for example, to the important part that the Young Farmers have continued to play at Buzet-sur-Tarn. To illustrate the decline of Cadours's organized farmers anecdotally: by the mid-1960s the opponents of land consolidation complained loudly about the claim by the supporters of *remembrement* that "the leaders of the farmers' union had been consulted and had endorsed the demand [for government-sponsored land consolidation]." Opponents of land consolidation rejoined that considering that there were fewer than half a dozen active union members at Cadours, this claim was blatant hypocrisy.

19. For a summary and analysis of the legal provisions involved, see *Code rural* (Paris: Dalloz, 1981), 351–62, 922–23, and Cotton, *Législation agricole*, 141–42, 517–22.

20. I plead guilty to what is almost certainly a gross oversimplification when I link changes in familial mores entirely to such questions of supply and demand, and, consequent changes in the bargaining power of wives and children. I am aware of the fact that the increasing exposure of rural families to national norms (which, on closer inspection, usually turn out to be middle-class norms) through more pervasive and influential media (first radio, now television) is likely to have contributed to changing standards of everyday behavior. On the basis of about one hundred interviews with French farmers and those in close contact with them, I am nonetheless convinced that I am not exaggerating by much. Even though I made no deliberate effort to explore this area, quite spontaneously the issue came up again and again in the course of my interviews.

21. Ville de Cadours, *Concours agricole cantonal, 9 et 10 séptembre 1967. Programme officiel* (Cadours, 1967).

22. Nicolas Peneff, *Histoire du foncier et du remembrement à Cadours* (Nantes: Presses de l'université de Nantes, 1980): 185.

23. Peneff, *Histoire*, 35.

24. Peneff, *Histoire*, 37–40.

25. By September 1967, for example, he was making public speeches at the annual cantonal fair in which land consolidation and irrigation were explicitly linked as the twin hopes of agricultural progress for Cadours, Ville de Cadours, *Concours agricole*.

26. One of the paradoxical results of the troubled *remembrement* at Cadours was that most neighboring *communes* chose to consolidate their farmland by less authoritarian methods, that is, by the use of multilateral exchanges, often sponsored by the Compagnie d'aménagement des Côteaux de Gascogne, the regional providers of irrigation. For a summary of and a commentary on the legislation on the French statute books, see Cotton, *Législation agricole*, 510–12.

27. Peneff, *Histoire*, 114–21.

28. The foregoing account is based on the official documentation in the files of the BDR, DDA-Toulouse, files that include the sometimes lengthy remonstrances by opponents of land consolidation. At least equally informative were the more or less extensive, indeed, sometimes interminable, comments by virtually every Cadours farmer that I interviewed. The already cited published monograph by the former mayor, while something of a brief for the defense, was also very informative.

29. "Minutes of the Municipal Council, Cadours," 19 April 1975, Cadours municipal archives. I also examined additional relevant documents dealing with the problem of the lake, both in the municipal archives and in private hands.

30. "Statistique agricole annuelle," 1913, ADHG M 299.

31. Fernand Roblin, "L'ail violet de Cadours," in Ville de Cadours, *Concours agricole*.

32. Cotton, *Législation agricole*, 444–46. The original laws of 1960 that sought to restrict what was considered the undue accumulation (*cumul*) of farmland by a single farmer have been weakened by subsequent amendments as well as by judicial interpretation. The laws are frequently unenforced and easy to circumvent. However, I did find one substantial farmer at Cadours against whom the law was successfully invoked, the only such instance in six *communes* studied.

Chapter 5
Where Have All the Sharecroppers Gone?

1. The contemporary section of this chapter, as well as some of the information from just prior to World War I onward, is based on extended interviews with the following farmers, technicians, and officials:

Farmers, retired farmers, present or former landed proprietors: Claude Bégard, Jean Bouchet, Joseph Deville, François Fontorbe, Pascal and Doria Fontorbe, Edouard and François Hébrard, François Manenc, Jean-Louis Marty, Fortuné Messal, Gerald and René Montserret, Hervé Pech, Charles-Louis de Thonel d'Orgeix, Guy de Thonel d'Orgeix, Roger Trantoul.

Officials, former officials, former farm stewards, agricultural technicians: Amams, Benet, Berthier, Canut, Pierre Deville, Miran, Mittou, Pradine.

2. For the legend connecting the village name with a she wolf, see Cassagne, "Monographie de Loubens," (1886) ADHG, Br. 4° 193. For what it is worth, to judge by King Henri II's letter patent of December 1556 granting the village a weekly market and a yearly fair, Loubens was at that time spelled "Lobenx" (Gounon C papers, document hand-copied from the National Archives, "Régistre du Trésor des Chartes," Côte 263, item no. 569.) I don't have the philological qualifications to say whether this invalidates the legend, but the French southwest place-names ending in "x" are reputed to have Gascon or Basque etymologies, which would make the "louve" derivation rather unlikely. The Gounon family papers had, prior to the time that I consulted them, been divided between two brothers, Guy d'Orgeix de Thonel and Charles-Louis d'Orgeix de Thonel, who are descendants of Joseph François de Gounon. The latter had been a cloth merchant who gained nobility by becoming *capitoul*, that is, alderman, of Toulouse in the third quarter of the eighteenth century. By purchasing the lands and

lordship of Loubens, Gounon added "de Loubens" to his name. I am grateful to his two descendants for opening their papers to me. Guy, as the elder, inherited what were perceived as the historic documents, many of which have to do with the acquisition of the lands and lordship of Loubens by Joseph de Gounon in 1771. These will be referred to as the Gounon G papers; the much more voluminous Gounon C papers were, when in 1983 I examined them, a vast, unsorted mass of miscellaneous documents two to three feet deep that filled one tower room of the château of Loubens, constituting part of the inheritance of Charles-Louis d'Orgeix de Thonel. After several days of sifting through these mounds of dusty materials dating from the Napoleonic period to the turn of the the twentieth century, I selected all of the estate records and account books that I discovered for copying. I understand that in the intervening years M. Charles d'Orgeix has turned over some of this cache of documents to the departmental archives of Haute-Garonne and organized the rest. For scanty biographical data on their ancestor who became Gounon de Loubens, see J. Villain, *La France moderne généalogique. Haute-Garonne et Ariège*, vol. 2 (Montpellier: Firmin, Montane, and Sicardi, 1911): 633–34, and Alphonse Brémond, *Nobiliaire toulousain* (Toulouse: Bonnal and Gibrac, 1863): 405.

3. The fact that the cadastre of 1731 makes mention of 150 houses without, however, indicating how many of these were occupied, suggests a figure of about six hundred inhabitants as a reasonable estimate.

4. I am basing my comments on the population figures below, which are derived as follows: for the eighteenth-century figures for the number of households (with the 1788 figure almost certainly too low), and the 1790 census, Frêche, "Dénombrement," 417. I have had occasion to examine the following *listes nominatives* in some detail: 1804, 1836 (not utilized because, for reasons unknown, it is short by about 140 inhabitants), 1896, 1911, 1926, 1936, 1946, 1962, and 1975. These may be found, respectively in the Loubens municipal archives for the first two, the rest in ADHG M 201, 235, 255, 925/3, 3455/7. The remaining figures are taken from the *Annuaire de la Haute-Garonne*.

Census by Household					
1713	146	1781	127	1788	110

Censuses or Yearly Counts by Inhabitants					
1790	690				
1804	681	1831	747	1842	749
1852	726	1861	634	1872	640
1881	571	1891	534	1896	482
1901	466	1911	386	1921	missing
1926	334	1936	298	1946	266
1954	281	1962	210	1977	189

5. Theoretically, it should be possible to obtain very precise information as to who immigrated to where, because, ever since the nineteenth century, a French law has been on the books that requires the authorities in the *commune* of a person's death to notify the mayor of the deceased's birthplace. The mayor so notified is then required to record the dead person's place and time of death in

the margin of the official birth entry kept in the municipal archives. Unfortunately, anyone working with these records soon becomes aware that the law has only been intermittently honored.

It is possible, of course, by what amounts to a simplified family reconstitution, simply to search for local death certificates for all whose births were recorded in the community, thus determining who did and who did not stay in a community. However, in a village in which sharecroppers and *maîtres-valets* were both numerous and geographically mobile, the absence of a Loubens certificate may simply indicate that a particular family took up another farm in another *commune*, a phenomenon that is sociologically totally different from joining the rural exodus.

Another, less cumbersome, technique is available when successive name-by-name and address-by-address census lists are available, as they are for Loubens beginning in 1896. (Two earlier lists for 1804 and 1836—incomplete for the latter in any event—are too isolated to be of use.) With the help of the local land register (cadastre), it is possible to ascertain who the owners of each farm (designated by a hamlet name or *lieu-dit*) were and consequently to distinguish resident proprietors from tenant farmers. This in turn permits the researcher to distinguish between genuine "emigration" and the usual game of musical chairs that tenant farmers and their masters played continually. To the extent that the *listes nominatives* accurately specify occupation, this technique makes a meaningful, though still incomplete, analysis of the rural exodus possible.

Yet, as already noted, because of the fragmentary enforcement of the law calling for notification of the deceased's home *commune*, this approach fails to reveal where those who left went. For the period since about 1914, it is possible to fill in most of this gap by recourse to the retentive memories of local old-timers.

6. M. F. Pariset, *Economie rurale, moeurs et usages du Lauragais (Aude et Haute-Garonne)* (Paris, 1867; reprint, Marseille: Lafitte Reprints, 1979): 26–27. For a discussion of the statistical relationship between rural depopulation in Haute-Garonne and the growth of the city of Toulouse, see [Louis] Théron de Montaugé, *L'agriculture et les classes rurales dans le pays toulousain depuis le milieu du dix-huitième siècle* (Paris: Librairie agricole de la Maison rustique, 1869): 422–26, who also emphasizes the railroad builders' demand for unskilled labor. My only evidence for immigration to the lower Languedoc is impressionistic: in the Loubens Etat Civil there are a number of entries of natives whose place of death is given as the *département* of Hérault, the heart of the Languedoc vineyard.

7. These remarks are based on the complete letterbooks, 1896—1927, of Jerôme Olognon, the farm steward of the château of Loubens, that is, of the old Gounon-Loubens estate, by then under multiple family ownership. The Olognon letterbooks and account ledgers, kindly given to me by his descendants, the Deville family and now deposited at the ADHG, provide an extraordinarily vivid and informative picture of Loubens' tenant farming during the first quarter of the twentieth century. These will hereafter be referred to as Olognon papers.

8. This compares with a national figure of 12.7 percent for the over sixty age group in 1901. By 1921 the national figure was 13.9 percent, in 1946 16.0 percent, in 1962 17.1 percent and, thanks to France's postwar population spurt, it has not risen since: the national figure for 1975 is also 17.1 percent. *Annuaire statistique de la France. Résumé retrospectif* (Paris: INSEE, 1966): 406. Roland

Pressat, "La population française; mortalité, natalité, migration, vieilissement," in *Actes du colloque national sur la démographie française, Paris: 23–25 June 1980* (Paris: Presses universitaires de France, 1981): 32. From having a little less than 50 percent more old people than the national norm, in more recent years the old have been overrepresented by almost 100 percent.

9. Pariset, *Economie rurale*, 19.

10. Frêche, *Toulouse*, 776–96.

11. Pariset, *Economie rurale*, 62–63.

12. Georges Jorre, *Le terrefort toulousain et lauragais. Histoire et géographie agraires* (Toulouse: Privat, 1971): 162–64; Roger Brunet, *Les campagnes toulousaines. Etude géographique* (Toulouse: Boisseau, 1966): 318–23. The tradition of bourgeois acquisition of agricultural land in the vicinity of Toulouse may sharpen the picture of noble landlords, many of them of relatively recent *roturier* origins, as hard-nosed, "businesslike" administrators presented in Robert Forster, *The Nobility of Toulouse in the Eighteenth Century: A Social and Economic Study* (Baltimore, Md.: Johns Hopkins University Press, 1960): 47–87.

13. For the central role of pastel, a dye plant, in the economy of the Toulouse region, particularly in the sixteenth century, see Philippe Wolff, ed., *Histoire de Toulouse* (Toulouse: Privat, 1974): 223–36. For the key role of the canal opened at the turn of the eighteenth century, see Frêche, *Toulouse*, 619–29, and for the grain trade prior to its opening, pp. 750–58.

14. Gounon G papers.

15. Pariset, *Economie rurale, passim,* takes this state of affairs for granted for the Lauragais of the 1860s, but also points out that, if anything, divisions through inheritance had diluted landlord power by that time. See also Brunet, *Les campagnes toulousaines,* 373–75, and Jorré, *Le terrefort,* 160–64, 174–77, although the latter does not distinguish the Lauragais from other areas of the *terrefort* where noble and bourgeois proprietors played a lesser role.

16. Although I never came across any document spelling out this arrangement in so many words, a careful reading of the minutes of the Loubens municipal council (complete back to the 1730s) persuaded me that such was the nature of the local administration. The only peasants selected were what were locally known as *ménagers,* that is, peasants owning enough land of their own to support a family. This criterion meant that the bulk of the peasant population, who were sharecroppers, *brassiers,* day laborers, and *maîtres-valets,* were completely excluded from local government, save as taxpayers.

17. "Dénombrement de la Seigneurie de Loubens," [1771?]; for a person-by-person breakdown of manorial dues, see "Etat des censives de la seigneurie de Loubens appartenant à M. Loubens-Gounon, seigneur du lieu, dépouillé par le sieur Jean Reynès, fermier," 15 July 1792, Gounon G papers.

18. Even before the French Revolution, among the Gounon de Loubens holdings, only the home farm La Borio was considered noble land exempt from the *taille.* After the Revolution had abolished fiscal distinctions between noble and nonnoble land, Gounon paid a land tax of 5,240 livres or 30 percent of the land tax burden shouldered by all Loubens property owners combined. The next highest taxpayer, a bourgeois absentee from the then episcopal city of Lavaur 15 miles north of Loubens, paid 1,179 livres. "Matrice de la contribution foncière," 1791, ADHG 5 E 177. The Gounons' real estate holdings were not confined to Loubens

and their town house in Toulouse. They also owned the château and several farms in Fourquevaux, some 12 miles distant from Loubens, where they resided more readily than at the dank and run-down manor house in Loubens. The family also had extensive land holdings in what was then the Comté de Foix and what is now the *département* of Ariège. One unexpected finding was that for at least two generations following their ennoblement, the Gounons continued, as their surviving ledgers testify, to be active and successful as wholesale cloth merchants in Toulouse, a derogation that prior to the Revolution might have led to the cancellation of their noble status for pursuing an ignoble occupation. Gounon G and C papers.

19. "Rôle de la capitation," 1790, ADHG C 1196. The breakdown for personal servants is derived from the "Rôle de la capitation," 1781, ADHG 5 E 177 and the "Matrice du rôle de la contribution foncière," 1791, ADHG 5 E 177.

It should be pointed out that the notion of "farm" (then as now, the regional term is *métairie*) shares with the English manor the dubious distinction of being a term without any clear meaning. Some of the farms of the Gounon estate were so large (48 hectares or 120 acres) as to require eight adult males as plowmen; others were more nearly the ideal size for a nuclear family to work, that is, 12–15 hectares or 30–40 acres. Two small farms could easily be run as one or a large farm split into two. Among the farms of the Gounon estate, Jacques Auriol and Faysset, as well as La Borio grande and La Borio petite, were lumped together when rented to an extended family. At other times they were split up between unrelated tenants, which accounts for the fact that the total number of "farms" in the Gounon estate kept changing at a time when its overall acreage remained constant. In short, a "farm" or *métairie* (like the English manor) was not only variable in size, but was frequently an unstable arrangement. Farms therefore tend to provide imprecise and unreliable clues as to the permanent divisions of the land into agricultural units.

20. In 1791, the distribution of the Loubens land tax load, and therefore, roughly, of the land if quality is taken into account, was as follows (note that by 1791 fiscal privileges had already been abolished):

3	Ex-nobles (from Toulouse, Caraman, Villeneuve)	30.6%
14	Local bourgeois (including adjoining *communes*)	19.4%
10	Local professionals (including adjoining *communes*)	5.6%
3	Local merchants (including adjoining *communes*)	4.8%
7	Absentee bourgeois and merchants	9.9%
12	Loubens *ménagers*	10.3%
8	*Ménagers* from adjoining townships (*forains*)	2.9%
44	Local artisans, shopkeepers, tradesmen	9.1%
45	*Brassiers*, unclassifiable, excluding prior categories	7.4%
		100.0%

Note that I have gone back to prerevolutionary tax rolls to identify *bourgeois* as the term was then used, that is, to describe a nonnoble person of property without an acknowledged profession or trade. As noted earlier, prerevolutionary documents invariably single out middle- or upper-class persons by having their name preceded by a title such as M[onsieur]., Sieur, Mme, Dame or D[emoise]lle. Merchants and professionals were included among people so singled out, but if they also had an occupation, it is invariably included. In the eighteenth century, Toulousain bourgeois status (in the French eighteenth-century sense of the term) was indicated either by an actual listing as *bourgeois* or by default, in which case a title of respect was provided but no occupation listed. In the case of women of middle-class status, things are murkier and labels rarer. I may therefore have inflated the "bourgeois" group by including in that category some carelessly identified women who may have been the widows, daughters, or sisters of professionals and merchants.

21. The figure of seventeen is purely fortuitous. It happened to be the number of Loubens land-taxpayers in 1791 whose property was assessed as having an income in excess of 200 livres. This paragraph summarizes a rather painstaking analysis of the following documents: the *compoix* of 1731 (the name by which land registers were known in the eighteenth-century French southwest), a masterpiece of calligraphy that I consulted in the Loubens municipal archives (the "duplicate" copy in the ADHG is incomplete and therefore useless); the Loubens "Matrice du rôle de la contribution foncière," 1791, ADHG 5 E 177; the Loubens "Matrice cadastrale," 1828–1913, ADHG, 1731/505; the Loubens "Matrice cadastrale," 1914–1947, ADHG, 1126W/346 (I specifically analyzed property distribution in 1914 and 1926); the Loubens "Matrice cadastrale," 1948–1969, all in the Loubens municipal archives.

22. For determining the number of sharecroppers in Loubens by the 1930s contemporary census lists are useless. With democratic abandon and a splendid disregard for status, the census of 1936 listed all farmers except day laborers and truck gardeners as *cultivateurs*. I have therefore relied on the official list of wheat producers for 1935–1936, which clearly indicates for each of the thirty-one names whether the farmer was a sharecropper or an owner and, if a sharecropper, who the land's owner was. "Etat détaillé par commune des déclarations des superficies ensemencées et des surfaces des terres labourables. Campagne agricole, 1935–36. Loubens," Loubens municipal archives.

23. My remarks in this and the paragraphs that follow rely on an examination of a number of eighteenth-century sharecropping contracts drawn from the files of the Loubens *notaire* and dealing with Loubens farms that were leased. In order to be able to utilize what would otherwise have been a forbidding number of documents, I sampled the first two of every fifteen years of notarial records, singling out only those records in which one of the parties involved resided in or was a property owner in Loubens. I did this for the years beginning with 1731–1732 and ending with 1911–1912, three years before Loubens's last notary retired.

The sharecropping contracts examined for this study are as follows:

1731–1732: seven contracts, ADHG, 3 E 20341 [Denegret, *notaire*], 35–36, 155–156A, 160–160A, 164–65, 173A–174, 175–175A, 183–183A.

1746–1747: twelve contracts, ADHG 3 E 20344 [Denegret, *notaire*], 243–244, 3 E 20345, 27–27A, 92A–93A, 109A–110A, 112–112A, 114–114A, 128–128A, 135–135A, 156–157, 167–168A, 229–230.

1763–1764: one contract, ADHG 3 E 20349 [Denegret, *notaire*], 195–96.

1777–1778: two contracts, ADHG 3 E 20356 [Boyssel, *notaire*], unpaginated two pages following p. 118, 148A–149A.

I found no sharecropping contracts in my nineteenth-century samples and only
one in my twentieth-century notarial selection. Incidentally, the Loubens notarial records dating from the mid-1860s to 1915 are in the custody of the official
successor to the *étude de notaire* at Loubens, namely the *notaire* at nearby Caraman, the *chef-lieu de canton*. The various nineteenth- and twentieth-century
tenancy agreements or contracts cited elsewhere turned up in the Gounon C, the
Selme, and the Olognon papers.

24. For the tithe and for the payment in kind to harvesters in the eighteenth
century (known locally as *l'escoussure* or *l'excoussure*) in Loubens as amounting
to one-tenth each: the tithe was split between the parish priest (who received 50
percent; the parish including not only Loubens, but adjoining Vendine and Fran-
carville, with the main church in the last of the these), the chapter of the arch-
episcopal cathedral of Saint Etienne in Toulouse (25 percent), and the monastic
establishment at Moissac (25 percent). See Darailh, parish priest at Loubens to
Archbishop of Toulouse, 12 November 1781, ADHG 1 G 536 for the fact that the
traditional collection of the tithe had simply been taken over by the landlord.
One sharecropping contract is worth quoting: "the tithe or tenth of the grain will
be collected by M. Boyer [the farm steward] or his representative in the field to
be transported to the threshing floor in order to check on the yield." In other
words, ninety years after the French Revolution that had professed to abolish the
tithe, the latter was alive and well in Loubens: only the beneficiary had changed.
"Bail à colonage, M. de Sévignac-Antoine Bals," 28 December 1879, Olognon
papers. For the *escoussure*, see "Vérification des droits seigneuriaux et biens-
fonds nobles de la terre de Loubens et de ses dépendances," [1771?], Gounon G
papers. According to Pariset, *Economie rurale*, 240, the custom of the harvesters
receiving one-tenth of the grain harvested was still in force as late as the mid-
1860s in much of the Lauragais, though Pariset seems to have been more familiar
with that part of the region within the Aude *département*. However, if a sharecropping contract dated 1 November 1974, between Victor Gounon-Loubens and
Jean Nougarlot concerning the farm La Maynade was typical of Loubens, by that
time the share of the harvesters had increased to one-eighth, rather than one-
tenth, of the grain harvested. Olognon papers.

The percentage set aside for seed was specified in "Vérification" cited above
and amounted to 29.4 percent of the average amount of wheat harvested and 36.7
percent of what was left after the harvesters had been paid off. However, this was
a document drawn up by a lawyer using a set formula, though I have been unable
to pin down the exact nature of that formula. When it comes to real life (as it can
be traced in some surviving estate accounts), since harvests fluctuated in a range
of one to two and a half from year to year, but the amount of seed sown did not,
the ratio of seed to preceding harvest obviously fluctuated also. What I find less
easy to explain is that even on the same farm the seed requirement varied considerably from field to field. In any event, whoever drafted the "Vérification"
seems to have been excessively conservative in his allowances for seed. In practice, the ratio seems to have been more nearly in the four- or five-to-one range,
that is to say, that for every four or five bushels of grain harvested (disregarding
any deductions for tithe and harvesters) one bushel was set aside as seed for the
following year. Gounon C papers.

25. This statement is based on an examination of farm inventories, including

tools and farm implements, to be found in the Gounon C papers and the papers of another local land-owning family that appeared on the Loubens scene during the first quarter of the nineteenth century, the Selme (who, in the course of time added "de La Glazière" to their name) who built (or came to reside in) the second château in Loubens.

The inventories unfortunately pose major linguistic obstacles, in that as many as one-half of the objects listed in the 1790 inventories cannot be clearly identified, despite recourse to a dozen eighteenth- and nineteenth-century dictionaries, standard and agricultural, in French and in Occitan/Provençal. Notwithstanding these semantic problems, the inventories give the impression that they describe fully equipped farms.

By the early twentieth century, this situation had changed in that sharecroppers were expected to come, not with their own work animals, which continued to be furnished by the landlord, but with most of their own equipment, often including such new-fangled machinery as reapers, reaper-binders, and mechanical hay-rakes. Olognon papers.

26. For the consultative role of the justice of the peace in twentieth-century landlord-tenant controversies, see, for example, Olognon to Marquise d'Orgeix, 12 February 1926 and 31 March 1926, "Letterbook," April 1921–April 1927, Olognon papers. The sharecropping contract referred to was also found in the Olognon papers.

27. Sharecropping contract, 1 November 1874, Victor de Gounon-Loubens and Jean Nougarlot, article 6. Olognon papers.

28. After the establishment of a permanent village school in 1833, the minutes of the Loubens municipal council provide an annual list of pupils exempt from paying tuition. As far as I have been able to determine, all of these were the children of village artisans and none of sharecroppers or, for that matter, landowning peasants. Loubens municipal archives.

29. For the Selme estate, see, for example, the nonnotarized 1836 sharecropping agreement for the *métairie* En Bousquet, the Selme home farm, and a 1912 contract for En Mariel. Selme papers. For the château de Loubens estate, a draft of a 1910 sharecropping contract for the farm of Enquit spells out that the sharecropper is "to obey his master or his representative any time that he is ordered to undertake repairs on the farm or any new construction for the master's personal use and to lend his arms or his cattle and do so without payment." Gounon C papers.

30. "Bail à colonage de Gounon-Loubens-Bélaval," 7 October 1911, no. 35, [Mazières, *notaire*].

31. The skirmishing began shortly after the end of World War I. Olognon to Comtesse de la Combe, 22 June 1920, "Letterbook," January 1917–April 1921; 23 April 1921, "Letterbook," April 1921–April 1927; Olognon to Marthe de Bataille de Sévignac (same correspondent, change of marital status) 5 May 1923, 11 April 1924, 26 June 1924; Olognon to Henri d'Orgeix, 14 March 1926; Olognon to Marquise d'Orgeix, 23 March 1926, "Letterbook," April 1921–April 1927, Olognon papers.

32. *Marthe* (San Diego: Harcourt, Brace, Jovanovich, 1984), *passim*. This fas-

cinating "nonfiction novel" is a collection of authentic letters, mostly from various members of one family of country nobles, spanning ten years around the turn of the twentieth century and focused on the family "problem child," a wayward young woman named Marthe. Even though the letters' editor (a descendant of the family) has chosen to remain anonymous and names and places are disguised, I believe it deserves to be taken seriously as an extraordinary source for the history of late-nineteenth century aristocratic *mentalités*.

33. "Bail à colonage, Jean-Victor Thuriès-Gabriel Colombiès et Pierre Escarboutel," 24 September 1868, Olognon papers.

34. "Vérification des droits seigneuriaux et biens-fonds nobles de la terre de Loubens et ses dépendances," [1771?], Gounon G papers, is very specific on the relative value of wheat and other crops. It is true that changes in agricultural techniques, particularly the growing importance of artificial, that is, sown, meadows and the decline of fallow in the ensuing fifty years changed the picture somewhat. The ratio between wheat and other grain crops probably was little changed until the the introduction of American hybrid corn after World War II, but the share of income from cattle sales almost certainly increased markedly in the course of the nineteenth century.

35. Suzanne Fiette, "Propriétaire et exploitants dans un grand domaine du Lauragais à la fin de l'ancien régime et au début de la Révolution," *Revue d'histoire moderne et contemporaine* 29 (1982): 177–213.

36. At least this is the suggestion of Frêche, *Toulouse*, 336–37.

37. Of the many tax lists extant for the 1780s and 1790s, the most comprehensive seems to be that of the "Contribution mobiliaire de 1791" (Loubens municipal archives), which lists twenty-seven sharecroppers (probably an undercount) as against a single *maître-valet* (almost certainly an undercount). Two or three additional individuals without occupational or status categories listed, but living on farms known to have been worked by tenants, are likely to have been *maîtres-valets*. The highest official listing is three *maîtres-valets* in the "Rôle du Vingtième," 1786, ADHG C 1196.

38. Untitled and informal, yet detailed, description and inventory of the Gounon de Loubens estate, [1772?], Gounon G papers.

39. Gounon C papers. At the time that Olognon assumed his stewardship in 1896, the last Gounon farm was being reconverted from *maître-valetage* to sharecropping. Olognon papers.

40. Pariset, *Economie rurale*, assumes throughout what is obviously a very well-informed analysis of the Lauragais in the 1860s that most estates in the region were worked by *maîtres-valets*. For example, on p. 87, Pariset states, "if the running of an estate rests on three kinds of participants—the landowner, the *maîtres-valets*, and the day laborers or harvesters—the second of these is most characteristic of the way that landed property is managed in the Lauragais." Sharecroppers or sharecropping get only passing mention. Indeed, the book is partly a tract against the *maître-valetage* system, with sharecropping or leasing farms for a fixed rent being held out as far more effective alternatives.

41. I examined the following documents or sources: "Mercuriales grains, 1698-An VI," Caraman municipal archives, AC 53. These series are a nightmare to use,

since they go by the local grain measure, namely by *sétiers de Caraman* (which differ from the *sétiers de Toulouse*), provide monthly prices without yearly averages, and specify the prices for three different grades of wheat without indicating the amounts of each bought and sold. The municipal archives of Toulouse, 4 F 27–28, have the *mercuriales des grains* from 1802 through 1866. Nineteenth-century grain-price listings are in hectoliters (the volume equivalents of 78–82 kilograms of wheat, but of only 50 kilograms of the lighter oats). Post–World War II prices are normally quoted in *quintaux*, that is in 100 kilograms. I have been unable to locate any systematic listing of later prices and neither, evidently have Georges and Geneviève Frêche, *Les prix des grains, des vins et des légumes à Toulouse (1468–1868)* (Paris: Presses universitaires de France, 1967). The *Annuaire de la Haute-Garonne et de l'Ariège* occasionally, but only occasionally, notes regional grain prices. A systematic use of the Gounon, Selme, and Olognon papers could establish the basis for a regional *mercuriale* for late–nineteenth and early–twentieth century wheat prices.

42. A draft of a sharecropping contract, dated 22 January 1899 (Bail à colonage, Louis-Jules Gounon-Loubens-Joseph Jaussely) and annotated by Gounon, specified in the margin that the tithe on oats had been abandoned six years earlier. Also Olognon to Mlle De Sévignac, 8 August 1915, "Letterbook," March 1915–September 1916, Olognon papers.

43. For a contract calling for two-thirds of the wheat crop going to the landowner, see "Bail à Colonage de M. Jules de Gounon et Auguste Alboui du 27 novembre 1875" and "Bail à Colonage de M. de Sévignac et Bals, Antoine du 28 décembre 1879"; another nonnotarized contract, dated 1 November 1874 (Victor de Gounon-Loubens and Jean Nougarlot) also called for a two-thirds/one-third division, but it was amended on 28 October 1878, changing the division to fifty-fifty after the landlord had collected his tithe. However as late as 1881, some of the Gounon tenants were still paying two-thirds of the wheat crop. "Bail à colonage, Jules de Gounon-Loubens-Marius Pradelles," 4 December 1881, Gounon C papers. For a similar two-thirds/one-third division of profits and losses on cattle, unnotarized contract, M. Selme and Jean and Paul Bousquet, undated, to go into effect 1 November 1836, *Métairie* En Bousquet, Selme papers.

44. For the farm steward's reaction to finding a replacement sharecropper prior to World War I, see Olognon to Mme de Gounon-Loubens and Mme la Marquise d'Orgeix, 4 January 1909, "Letterbook," 25 July 1908–8 October 1909; for an example of the latter, see Olognon to Baron de Puymaurins, 16 January 1923, *Letterbook*, 23 April 1921–22 April 1927, Olognon papers.

45. "Etat détaillé par commune des déclarations des superficies ensemencées en blé et des surfaces des terres labourables. Campagne agricole 1935–36. Loubens," Loubens municipal archives.

46. "Vérification des droits seigneuriaux et biens fonds nobles de la terre de Loubens et ses dependances pour tirer du Produit ce qu'elle peut etre cottizée[!] de 20me noble," undated [1771?], Gounon G papers.

47. Anyone dealing with such problems of agricultural change is evidently indebted to Marc Bloch's thesis of the cattle-manure-fodder bottleneck as the great obstacle to agricultural progress. Unfortunately the Selme and Gounon C papers

are too fragmentary to determine quantitatively just what effects the conquest of the fallow through fodder crops had on Loubens agriculture.

48. The theories and practice of eighteenth-century French agriculture are treated in exhaustive detail in André J. Bourde, *Agronomie et agronomes en France au XVIIIe siècle*, 3 parts. (Paris: SEVPEN, 1967). For general agricultural practice, see pt. 2, 457–508, 597–617; for specific foreign examples and the extent to which they were known, see pt. 1, 277–310, 425–432. For eighteenth-century *regional* agricultural practice and ideas, see Frêche, *Toulouse*, 231–87.

49. Inventory of the *métairie* La Borio, 1771, Gounon G papers.

50. "Vérification," Gounon G papers.

51. Gounon de Loubens accounts, 1 July 1771–30 June 1772, Gounon G papers.

52. "Compte annuel rendu par M. le sous-préfet de Villefranche concernant l'état de l'agriculture dans son arrondissement," 1816, ADHG 11 M 7.

53. "Contrat de colonage, Jean-Victor Thuries-Gabriel Colombiès et Pierre Escarboutel," 24 September 1868. "Police de métayage, Victor de Gounon-Loubens-Jean Nougarlot," 1 November 1874; "Bail à colonage, Jules de Gounon-Auguste Alboui," 27 November, 1875; "Bail à colonage, Auguste Hébrard-Charles Sablayrolles," 22 January 1882. Olognon papers.

54. Pariset, *Economie rurale*, 55–57.

55. "Questionnaire de la statistique agricole décennale de 1892. Loubens," ADHG M 3.

56. These generalizations are based on a careful perusal and analysis of the accounts (as distinct from the related correspondence) of half a dozen farms supervised by Olognon, which can be followed from 1896–1898 to the late 1930s. Olognon papers.

57. Generalizations based on the "Letterbooks", 1896–1927, Olognon papers.

58. Olognon to housekeeper of Jules de Gounon, 18 August 1915, "Letterbook," 26 March 1915–30 December 1916, Olognon papers.

59. "Letterbooks," *passim*, Olognon papers.

60. There is no indication in the thirty years of Olognon's correspondence with his employers that he and they ever participated in the purchase of farm equipment on the part of their sharecroppers. Olognon papers, *passim*.

61. Gounon C papers.

62. Gounon C papers.

63. On the yearly accounts of both the Gounon and Selme *métairies* the fee owed to the harvesting contractor always figures prominently, Gounon C and Selme papers. As to the contractor covering more than one *commune*, this was the information I was given by those who remembered the old threshing machines.

64. The changeover can be traced in Olognon's account books for the farms under his supervision, 1896 to 1901, where the steam engine shows up as an expenditure for coal. The same accounts also have an item for the *trieuse*, the use of the grain sorter. The resistance to the engine-powered thresher may be seen in Olognon to M. de Sévignac, 2 July 1897, and its acceptance by another Gounon heir in Olognon to M. Jules de Gounon-Loubens, 11 July 1897, "Letter-

book," 3 April 1897–28 November 1897. Olognon himself supported switching to the more rapid and efficient engine-driven machine on the grounds that it did a lot to save wear and tear on the oxen. Olognon papers. Evidence for the regular purchase of copper sulphate for dipping seed grain may be found in both the Gounon C and the Selme papers.

65. "Enquête agricole de 1929. Economie rurale," Loubens-Lauragais, ADHG M 1022.

66. "Enquête agricole de 1929."

67. "Statistique agricole annuelle," 1932, Loubens-Lauragais, ADHG M 1073.

68. Duby and Wallon, *Histoire de la France rurale*, vol. 4, 574–75.

69. The juridical foundation of this revolution is complicated in itself, but it also continues to be modified by new legislation, judicial review, and local custom. For the basic legislative texts, see *Code rural et code forestier* (Paris: Dalloz, 1981), 231–72; for the text with legal annotations and/or commentary, see *Code des loyers et de la co-propriété*, 2d ed. (Paris: Dalloz, 1972) and Robert Poirel, *Manuel des baux ruraux* (Paris: Dalloz, 1971), *passim*. For the layman, a much clearer, though still technical, summary may be found in Cotton, *Législation agricole*, 315–410. For an analysis as to whether and to what extent French legislation doomed sharecropping in Loubens, see Peter H. Amann, "French Sharecropping Revisited: The Case of the Lauragais," to appear in the July or October 1990 issue of *European History Quarterly*, particularly the last part of the article.

70. In Loubens this happens to be the only such case. However, I was told of an aristocratic maiden lady in an adjoining township who continued to run six or eight *métairies* with no concession whatever to the twentieth century.

71. How adequately depends entirely on the accounting procedures followed. If the net income was figured as a proportion of the then current (1982) market price for farmland (about Fr 30,000 per hectare for prime agricultural land in Loubens), the return was unlikely to exceed 4 percent on the investment. If, on the other hand, the return was figured on the actual cost of land purchased in the early 1950s (but adjusted for inflation), the rate of return would be three to four times as high.

72. "RGA inventaires préliminaires, fiche communale Loubens-Lauragais," 1979, DDA-Toulouse, BDS. According to this survey, of a total agricultural acreage of 469 hectares, 185 were farmed by owner-operators, 174 by officially registered leaseholders, and 93 by sharecroppers, which leaves 16 hectares unaccounted for and presumably rented out "under the counter." Unfortunately these survey results are suspect, because all earlier surveys listed the agricultural land within the *commune* of Loubens-Lauragais as consisting of 616 hectares and *not* 469 hectares. The fact that one-quarter of the township's land is unaccounted for makes this survey virtually useless.

73. I am including the smaller of the two surviving sharecropped farms in Loubens, since with 32 hectares leased on half shares, in terms of net income, that farm was the equivalent of a farm of 16 hectares free and clear. In real life, the life-style and attitudes of the sharecroppers occupying that farm were undistinguishable from those of the rest of the "survivors."

74. *Remembrement* was requested by the Loubens municipal council in February 1968 and officially authorized by prefectoral decree eight months later. It was then held up for five years for lack of funding. The actual land consolidation and related public works were completed with a minimum of fuss in only three and one-half years, between May 1973 and November 1976. Loubens-Lauragais dossier, DDA-Toulouse, BDR.

75. Since Marcel Tillet acquired his farm through the retirement and not the death of his father, technically the son became the owner by a juridically specified form of parental gift called *donation-partage*. In the division of the property, his only advantage over his siblings would have been a hypothetical "wage" credit allocated to the heir taking over a farm, provided the latter had previously been an unpaid "family helper" (*aide familial*) on his parents' farm. Even so, this "extra credit" was reckoned at only one-half of the average wage of a farm hand receiving room and board up to a maximum of ten years. Cotton, *Législation agricole*, 69–99.

Chapter 6
Historical Perspectives

1. Since land consolidation is a highly technical undertaking, yet the people writing about it tend to be both technicians *and* true believers, the fervor of the introduction usually contrasts with the sobriety of the body of such works. The works cited as examples date, respectively, from 1933, 1959, and 1980, but the tone is very similar and almost any book on the subject could have been cited. M. E. Damuzeaux, *Le remembrement de la propriété foncière dans les Ardennes* (Sedan: Imprimerie Suzaine, 1933); Le Génie rural, La Chambre d'Agriculture, *Remembrements en Meuse* (Bar-le-Duc, 1959); Christian Atias and Didier Linotte, *Le remembrement rural* (Paris: Librairies techniques, 1980).

2. I encountered a case in point in studying agricultural modernization in the *commune* of Carbonne, located largely in the valley of the Garonne river, about 30 miles south of Toulouse. Carbonne was a community where practically all farmers with farms in the plain had begun to irrigate, taking advantage of a vast aquifer very close to the surface. Carbonne was also a township where *remembrement* had been begun, but where it was thwarted by local opposition. The authorities retaliated by retaining the ban on land transfers that had been in effect during land consolidation, a ban that remained in force for the ensuing ten years that happened to coincide with the diffusion of irrigated agriculture in the township. Unable to effect legal field exchanges and unable to irrigate amidst enclaved fields, Carbonne farmers solved their problem by informal, indeed verbal, bi- or trilateral swaps which had no standing in law (and posed potential problems in case of inheritance or sale), but which were apparently quite effective in solving the technical problems involved in adapting the Carbonne field system to the demands of irrigation.

3. In France's First Plan, initiated in 1946 under Jean Monnet, agricultural machinery was one of the six high-priorities items and called for ultimate investments of Fr 10 billion, Economic Cooperation Administration, *European Recovery Program. France. Country Study.* (Washington, D.C.: U.S. Government

Printing Office, 1949): 72. Yves Ullmo, *La planification en France* (Paris: Dalloz, 1974): 5–7; Pierre Bauchet, *La planification française* (Paris: Editione du Seuil, 1966): 91–92. By far the fullest account of French official policies on every aspect of agriculture may be found in Warren C. Baum, *The French Economy and the State* (Princeton, N.J.: Princeton University Press, 1958): 21–24, 285–314, but from a different perspective, Pierre Barral, *Les agrariens français de Méline à Pisani* (Paris: Armand Colin, 1968): 310–24.

4. Between 1947 and 1953, 45 percent of all tractors sold in France were built abroad, most of them in Britain and Germany, both countries, of course also linked to France in the European Recovery Program. Baum, *The French Economy*, 297–98.

5. Hypothetical retrospective balance sheets are, admittedly, subject to considerable and unverifiable error, particularly since in 1952 continuing peasant family autoconsumption made *market* farm prices less than a central concern. However, assuming a typical Lauragais farm of 12 hectares (but *untypically* owner-operated and therefore *not* subject to the usual fifty-fifty split between owner and tenant), I also assumed a triennial rotation of wheat, corn, and fodder, with a pair of work-oxen and four cows, yielding three calves of 225 kilograms per year. The wheat and corn crops alone (optimistically figuring 2 tons of both wheat and corn per hectare) would have grossed the equivalent of twice the yearly average wage of male industrial workers and it is not unreasonable to postulate that the sale of calves would have covered all farm production costs including real estate taxes. Any analysis of a farm operation in 1982 terms gets much more complicated because of greatly increased production costs. I have therefore assumed (again optimistically) a figure of a net yield of Fr 2,000 (close to the optimum and probably never obtained on that small a farm) for both wheat and (nonirrigated) corn and a doubling of the calf production to six calves a year. One problem with these assumptions is the fact that modern farm machinery cannot be amortized on that small a farm, even though the soil-packing effect of modern harvesters makes older and lighter equipment (which such a small farm *might* be able to support financially) increasingly impractical. I have also omitted the raising of fowl and pigs, both of which would, of course, cut into the quantity of grain marketed, while adding to the marketable product. In short, if I have conveyed these changes in rough approximations rather than in neat percentages, I have done so because more precise figures would be patently fake.

I have been able to reconstruct a price series for calves (assuming, on the basis of my interviews, that farmers get about two-thirds of the price per kilogram quoted on the Paris wholesale market for first quality veal-on-the-hoof), wheat, and corn from the *Annuaire statistique de la France* and have, as noted earlier, used a conversion table published as "Le pouvoir d'achat du Franc, de 1914 à 1979," *Chambres d'agriculture* 51 (1980), supplement to no. 662–63, brought up to 1982 by adjusting for reported inflation rates for 1979–1982. Annual average (hourly) industrial wages in France are derived from the *International Labour Office Yearbooks*. In order to get weekly wages, I have multiplied by forty-six hours for 1952, by thirty-nine for 1982. To obtain annual incomes, I have multiplied weekly wages by fifty.

6. For a quick sketch as to how the French social security system for farmers works and how it got that way, see Roger Jambu-Merlin, *La sécurité sociale* (Paris: Armand Colin, 1970): 52–53. Not only is each regional or departmental *caisse* quite autonomous, but it is in turn subdivided into semiautonomous *sous-caisses*, each dealing with its specific concern, such as family allocation, insurance against sickness, old-age pensions, and so forth. I could find no publication summarizing the current status of farmers' social security dues in Haute-Garonne. Presumably it would take a close analysis of yearly reports, preferably by an accountant, to determine just how much farmers end up paying in a given year and on what basis. Since the issue is at best marginal to this study, I did not pursue the matter further.

7. I used the base-year 1981, as being the year closest to the period of my interviews in that 1981 was the latest year for which I could find average wholesale corn *prices* (as against *price indices* to which the editors of the *Annuaire statistique de la France* chose to switch in 1982). I assumed that French average wholesale grain prices would approximate the official support prices for these farm products by the EEC I might note that if I failed in my attempt to obtain authoritative information on prices from the EEC, it was not from lack of trying: its official publications are generous with indices, percentage increases, and yearly conversion figures from the European écu to national currencies, but amazingly secretive as to actual guaranteed prices for specific farm products. A direct letter of inquiry evoked an evasive bureaucratic response. For world prices, I used the export prices for wheat and corn furnished by *Statistical Yearbook of the United States* (Washington, D.C.: U.S. Government Printing Office, 1987) published by the U.S. Bureau of the Census, converting bushels to hectoliters and using a conversion factor of 1.25 to get *quintaux* (assuming 80 kilograms to the hectoliter, which is fairly accurate for wheat; corn, even when marketed as "dry," varies in weight). The average U.S. dollar to French franc exchange rate for 1981 (Fr 5.41 to $1.00) was also derived from the same U.S. source. In 1981 the world price for wheat per one hundred kilograms was $93.32 and for corn $74.13; the average French wholesale price (and, presumably, approximate EEC support price) was $113.30 for wheat, $116.53 for corn. The best, or at least most readable, brief account of the functioning of French agriculture within the current EEC framework may be found in John Ardagh, *France Today* (London, 1988), 213–18.

Bibliography

Contents

Oral Sources

The names and status of the people who spoke to me either about their experience as farmers or in connection with their acquaintance with one or more given communities is indicated in the first footnote to each "village" chapter. Because I promised privacy to my informants, I have deliberately failed to attribute information derived from oral sources to specific and identifiable informants. I have, of course, made every effort not to go beyond the evidence provided by the taped interviews. For the same reasons of respecting the privacy of my respondents, I have changed the names of informants and of other local inhabitants still living (in 1982–1983) that make their appearance in this book. Having admittedly tampered with names, I have, however, avoided "doctoring" the descriptions of my informants' views and/or personalities by creating unrecognizable composites.

Manuscript Sources

Like most historians, I went through a lot of archival materials that proved to be irrelevant or tangential to the study completed. I therefore see no useful purpose in listing what turned out to be sidetracks, dead ends, or even inspiration for future research projects. The same applies to archives I consulted, where visits to these archives—those of the Bureau du remembrement of the Ministry of Agriculture in Paris, of the Archdiocese of Toulouse, of the departments of Gers and of Tarn come to mind—made no substantial contribution to the writing of this book in its present form. Since I gave some details as to the nature of privately held source material utilized in this study in my notes, there is no need to do more here than briefly summarize the content of such collections. Where I relied on public depositories, or, as in the case of the Direction départementale de l'agriculture, working parts of what the French call *l'Administration*, I shall briefly summarize their respective contributions to this study.

Privately Held (as of 1983) Manuscript Sources

Copies or originals in possession of the author have been turned over to the ADHG in 1989 and are inventoried as Fonds Amann-Loubeus, sous série 32J.

Fontorbe papers. The accounts, beginning in 1909 and covering some fifty years, of a family of Loubens sharecroppers. Pascal Fontorbe, Loubens.

Gounon C papers. Nineteenth- and twentieth-century estate accounts, Loubens-Lauragais. Charles-Louis d'Orgeix de Thonel, Loubens and Toulouse.

Gounon G Papers. Miscellaneous documents related to the purchase and operation of the Gounon estate in eighteenth-century Loubens. Guy d'Orgeix de Thonel, Loubens.

Miran papers. The accounts of the former farm steward of the De La Tour estate in Loubens, 1940s–1960s. Joseph Miran, Loubens.

Olognon papers. The letterbooks and accounts of the farm steward of the Gounon estates (some seven *métairies* by then under multiple ownership), 1896-1935. Formerly in the possession of Joseph Deville, Loubens. Given to the author.

Selme papers. Miscellaneous estate accounts, 1820s–1920s, in the possession of Pierre Selme de la Glazière, Loubens.

French National Archives

I tried to find the cantonwide local appraisal of agricultural conditions in the files of the Enquête sur le travail agricole et industriel (1848), to the best of my knowledge the only local responses from agricultural inquiries that are to be found in the French National Archives. For the five cantons in which the *communes* studied here are located, only the responses to the inquiry from the canton of Aspet (Juzet d'Izaut) have been preserved (C. 953).

Departmental Archives of Haute-Garonne (ADHG)

Even though the departmental archives have by far the richest document collection bearing on the agricultural and demographic past of rural townships in Haute-Garonne, there are also frustrating gaps, either because such things as nineteenth-century census lists were discarded along the way, or, as I suspect in other instances (like missing local questionnaires to nineteenth-century decennial agricultural inquiries), the documents have never been classified. A major fire during World War II also had an impact. The major series that proved most useful were:

Br. is the classification for the individual township monographs written by village schoolmasters in or about 1885 on the instruction of the Ministry of Education.

Serie C contains most of the tax assessment lists and is most useful for the analysis of eighteenth-century communities. Some documents have never been classified.

Serie E shares tax assessments lists with Serie C.

Serie 3 E had all of the Loubens notarial records to 1864, which I supple-

mented with an examination of later records in the office of the notary of Caraman, the legal heir of the *étude* of Loubens.

Serie M has both yearly agricultural reports and local responses to agricultural
inquiries of various sorts (although the majority are lost) and all remaining demographic data, including detailed census lists (scarce prior to 1896).

ADHG also has the duplicates of local land registers (cadastres) kept in municipal archives, with cartons numbered and irregularly bearing a "W." Occasionally, as in the case of the *compoix* of 1731 for Loubens, cadastres kept in the
ADHG are fragmentary.

Municipal Archives

I systematically went through the local archives at Buzet, Saint Cézert, Juzet,
and Loubens. Typically, besides the required cadastres and, of course, the minutes of the deliberations of the municipal councils (which I systematically perused
from 1730 on only for Loubens), most municipal archives had some useful documents not found in ADHG that included early census lists, tax lists, and, for the
interwar period, lists of wheat growers. In the case of Cadours, which had more
extensive archives and a less casual atmosphere, my perusal was more selective,
although official correspondence dealing with the projected lake proved invaluable. At Caraman, the cantonal seat for the the canton in which Loubens is located, I merely consulted eighteenth- and nineteenth-century *mercuriales*, that
is, the series of market grain price reports (Serie AC). I did the same for Toulouse
(Serie 4 F).

Archives of the Direction Departementale de L'Agriculture Statistical Bureau

For the five *communes* that are the subject of this book, the Statistical Bureau
made available to me the following agricultural surveys: 1954–1955, 1970–1971,
and 1979.

Bureau du Remembrement

I was given full cooperation by the staff and allowed to peruse and copy all of
the documents relating to land consolidation in all of the five *communes* with
which I was concerned, usually a voluminous file that included not only the minutes of the land consolidation commission but also official recommendations and
appraisals, individual and collective petitions and remonstrances and their disposal at various levels, "before-and-after" detailed maps, and so forth. I was also
allowed to gather a dossier on the overall policies pursued in Haute-Garonne by
the Ministry of Agriculture, though little of this has found its way into this study.

Printed and Mimeographed Sources

In one way or another, almost every article and book that has ever dealt with
rural France may be thought of as relevant to this study. Yet given my grass roots

approach to agricultural modernization, my debt to many authors is indirect. As such, it is also often awkward to document. This bibliographical listing is therefore restricted to those works fruitfully consulted in preparing this book, those cited in the endnotes, or those alluded to in the text. Where a work first published in French is available in an English translation, I cite the English language version, with its date of publication, rather than the French language edition.

Agulhon, Maurice. *The Republic in the Village: The People of the Var from the French Revolution to the Second Republic.* Cambridge and New York: Cambridge University Press, 1982.

———. *The Republican Experiment, the Second Republic, 1848–1852.* Cambridge and New York: Cambridge University Press, 1983.

Annuaire de la Haute-Garonne et de l'Ariège. Toulouse: various printers, annual.

Annuaire statistique de la France. Résumé retrospectif. Paris: INSEE, 1966.

John Ardagh. *France Today.* London: Penguin Books, 1988.

———. *Rural France.* London: Century, 1983.

Atias Christian, and Didier Linotte. *Le remembrement rural.* Paris: Librairies techniques, 1980.

Aubin, Edmond. *Des abornements généraux et des remembrements. Opérations préalables à l'établissement des livres fonciers en France.* Paris: L. Boyer, 1902.

Barral, Pierre. *Les agrariens français de Méline à Pisani.* Paris: Armand Colin, 1968.

Bauchet, Pierre. *La planification française.* Paris: Editions du Seuil, 1966.

Baum, Warren C. *The French Economy and the State.* Princeton, N.J.: Princeton University Press, 1958.

Bloch, Marc. *French Rural History: An Essay on Its Basic Characteristics.* Berkeley: University of California Press, 1966.

Bourde, André J. *Agronomie et agronomes en France au XVIIIe siècle.* 3 parts. Paris: SEVPEN, 1967.

Braudel Fernand, and Ernest Labrousse, eds. *Histoire économique et sociale de la France.* Vol. 1, pt. 2, *Paysannerie et croissance,* by Emmanuel Le Roy Ladurie and Michel Morineau; Vol. 2, *Les derniers temps de l'âge seigneurial aux préludes de l'âge industriel (1660-1789),* by Ernest Labrousse, Pierre Léon, Pierre Goubert, Jean Bouvier, Charles Carrière, and Paul Harsin. Paris: Presses universitaires de France, 1977, 1970.

Brémond, Alphonse. *Nobiliaire toulousain.* Toulouse: Bonnal and Gibrac, 1863.

Brun, André. *Essai d'analyse d'une population agricole. Le canton de Caraman en Lauragais.* [Toulouse]: Institut national de la recherche agronomique, Association Midi-Pyrenées d'économie rurale, Toulouse and Direction des services agricoles de la Haute-Garonne, May 1963.

Brunet, Roger. *Les campagnes toulousaines. Etude géographique.* Toulouse: Boisseau, 1965.

Cepède, Michel. *Agriculture et alimentation en France durant la IIe Guerre mondiale.* Paris: Editions Génin, [1972?].

Chaianov, A. V. *The Theory of Peasant Economy.* Homewood, Ill.: Irwin, 1966.

Chambre de commerce et d'industrie de Paris. *Statistique de l'industrie à Paris résultant de l'enquête faite par la Chambre de commerce pour les années 1847–1848*. Paris: Guillaumin, 1851.

Chombart de Lauwe, Jean. *Bretagne et Pays de Garonne*. Paris: Presses universitaires de France, 1946.

Clarenc, Louis. "Les troubles de Barousse en 1848." *Annales du Midi* 35 (1957): 167–97.

Commission du Bilan. *La France en mai 1981*. Vol. 2, *Les activités productives*. Paris: La documentation française, 1981.

Coquerelle, Suzanne. "Les droits collectifs et les troubles agraires dans les Pyrenées (1848)." *Actes du 78e Congrès des sociétés savantes*, 354–64. Paris: Presses universitaires de France, 1954.

Cotton, Guy. *La Législation agricole*. Paris: Dalloz, 1975.

———. "Quelques réflexions sur les résultats de l'activité des S.A.F.E.R." *Information agricole*, No. 443, January 1974.

Damuzeaux, M. E. *Le remembrement de la propriété foncière dans les Ardennes*. Sedan: Imprimerie Suzaine, 1933.

Dequiral, René. "Essai sur les conditions économiques, sociales et démographiques des pays de Garonne au XIXe siècle." *Revue d'histoire économique et sociale* 29 (1951): 227–51.

Direction départementale de l'agriculture de la Haute-Garonne. "Plan d'aménagement rural—Canton d'Aspet. Etude des aspects fonciers et agricoles." Toulouse: DDA and Chambre d'agriculture de la Haute-Garonne, August 1982.

———. *Plan d'aménagement rural du Lauragais. Cantons de Carman—Lanta—Verfeil*. [Toulouse]: DDA and Chambre d'agriculture de la Haute-Garonne, December 1980.

———. *Plan d'aménagement rural du secteur de Villemur sur Tarn*. [Toulouse]: DDA and Chambre d'agriculture de la Haute-Garonne, December 1978.

Domergue, Achille. *Métrologie du département de la Haute-Garonne*. Toulouse: Lebon, 1839.

Duby Georges, and Armand Wallon, eds. *Histoire de la France rurale*. Paris: Editions du Seuil, 1975, 1976. 2, *L'Age classique des paysans, 1340–1789*, by Hughes Neveux, Jean Jacquart, and Emmanuel Le Roy Ladurie; Vol. 3, *Apogée et crise de la civilisation paysanne, 1789–1914*, by Maurice Agulhon, Gabriel Désert, and Robert Specklin; Vol. 4, *La fin de la France paysanne de 1914 à nos jours*, by Michel Gervais, Marcel Jollivet, and Yves Tavernier.

Dumas, André. *Le remembrement rural*. Paris: Sirey, 1963.

Dutil, Léon. *La Haute-Garonne et sa région. Géographie historique*. 2 vols. Toulouse and Paris: Privat and Didier, 1929.

Economic Cooperation Administration. *European Recovery Program. France. Country Study*. Washington, D.C.: U.S. Government Printing Office, 1949.

Fédération nationale des S.A.F.E.R. *S.A.F.E.R., organisation, fonctionnement*. Paris: Fédération nationale des S.A.F.E.R., 1979.

Fiette, Suzanne. "Propriétaire et exploitants dans un grand domaine du Lauragais à la fin de l'ancien régime et au début de la Révolution." *Revue d'histoire moderne et contemporaine* 29 (1982): 177–213.

Forster, Robert. *The Nobility of Toulouse in the Eighteenth Century: A Social and Economic Study.* Baltimore, Md.: Johns Hopkins University Press, 1960.

France. *Code des loyers et de la co-propriété,* 2d ed. Paris: Dalloz, 1972.

———. *Code rural et code forestier.* Paris: Dalloz, 1981.

———. *Recueil des textes relatifs au remembrement rural.* Paris: *Journal officiel de la République française,* 1980.

Frêche, Georges. "Dénombrement des feux et d'habitants de 2,973 communautés de la région toulousaine." *Annales de démographie historique* 4 (1969): 393–471.

———. *Toulouse et la région Midi-Pyrenées au siècle des lumières, vers 1670–1789.* [Toulouse?]: Editions Cujas, 1974.

Frêche, Georges, and Geneviève Frêche. *Les prix des grains, des vins et des légumes à Toulouse (1468–1868).* Paris: Presses universitaires de France, 1967.

Le Génie rural, La Chambre d'Agriculture. *Remembrements en Meuse.* Bar-le-duc, 1959.

Houée, Paul. *Les étapes du développement rural.* Vol. 1, *Une longue évolution (1815–1950);* Vol. 2, *La Révolution contemporaine (1950–1970).* Paris: Editions Economie et Humanisme, Les Editons ouvrières, 1972.

Houssel, J.-P., ed.; J.-C. Bonnet, S. Dontenwill, R. Estier, and P. Goujon. *Histoire des paysans du XVIIIe siècle à nos jours.* Roanne: Editions Horvath, [1976].

International Labour Office Yearbook. Geneva: International Labour Office, annual.

Jambu-Merlin, Roger. *La sécurité sociale.* Paris: Armand Colin, 1970.

Jollivet M., and H. Mendras, eds. Groupe de sociologie rurale. *Les collectivités rurales françaises, études comparatives de changement social.* Paris: A. Colin, 1971–.

Jorre, Georges. *Le terrefort toulousain et lauragais. Histoire et géographie agraires.* Toulouse: Privat, 1971.

Lecouteux, Edouard. *Guide du cultivateur améliorateur.* Paris: Dusacq, 1854.

Loubère, Leo. *The Red and the White: A History of Wine in France and Italy in the Nineteenth Century.* Albany: State University of New York Press, 1978.

Mahillon P., and M. Vinchent. *Etudes sur le Remembrement Rural.* Bruxelles: Larcier, 1955.

Margadant, Ted. *French Peasants in Revolt: The Insurrection of 1851.* Princeton, N.J.: Princeton University Press, 1980.

Marthe. San Diego: Harcourt, Brace, Jovanovich, 1984.

Mathieu de Dombasle, C.-J. A. *Calendrier du bon cultivateur ou Manuel de l'agriculteur practicien.* Paris: Huzard, 1830.

Mendras, Henri. *The Vanishing Peasant: Innovation and Change in the French Countryside.* Cambridge, Mass.: M.I.T. Press, [1970].

Mendras, Henri, and Ioan Mihailescu, eds. *Theories and Methods in Rural Community Studies.* Oxford and New York: Oxford University Press, 1982.

Merriman, John. *The Agony of the Republic.* New Haven, Conn.: Yale University Press, 1978.

————. Review of Eugen Weber, *Peasants into Frenchmen*. *Journal of Modern History*, 50 (1978): 534–36.

Pariset, M. F. *Economie rurale, moeurs et usages du Lauragais (Aude et Haute-Garonne)*. Paris: Bouchard-Huzard, 1867. Reprint. Marseille: Lafitte Reprints, 1979.

Pène, Jean. "La révolte de Barousse en 1848." *Revue de Comminges*, 68 (1955): 13–30, 79–91.

Peneff, Nicolas. *Histoire du foncier et du remembrement à Cadours*. Nantes: Presses de l'université de Nantes, 1980.

Poirel, Robert. *Manuel des baux ruraux*. Paris: Dalloz, 1971.

"Le pouvoir d'achat du Franc, de 1914 à 1979." *Chambres d'agriculture* 51 (1980), supplement to no. 662–63.

Pressat, Roland. "La population française; mortalité, natalité, migration, vieillissement." In *Actes du colloque national sur la démographie française, Paris, 23–25 June 1980*. Paris: Presses universitaires de France, 1981.

Roupnel, Gaston. *Histoire de la campagne française*, 2d ed. Paris: Plon, 1974.

Sabatié, F. "Buzet-sur-Tarn (1737–1792). Evolution démographique, économique et sociale." Toulouse: Mémoire de D.E.S., lettres, 1967.

Schmerber, J. Miquel. *La reorganisation foncière en France. Le rembrement rural*. Paris: Maison rustique, 1949.

Tables de comparaison entre les mésures anciennes et celles qui les remplacent. Toulouse: Douladoure, An X [1802].

Théron de Montaugé, [Louis]. *L'agriculture et les classes rurales dans le pays toulousain depuis le milieu du dix-huitième siècle*. Paris: Librairie agricole de la Maison rustique, 1869.

Thompson, E. P. *The Making of the English Working Class*. New York: Vintage Books, [1963].

Ufferte, Bernard. *Notices sur Cadours et sa région—son passé, ses usages, son développement*. Poitiers: Société française d'imprimerie, 1935.

Ullmo, Yves. *La planification en France*. Paris: Dalloz, 1974.

United States Bureau of the Census. *Statistical Yearbook of the United States*. Washington, D.C.: U.S. Government Printing Office, 1987.

Vallery-Radot, Maurice. *Remembrement rural et jurisprudence du Conseil d'Etat*, 2d ed. Coutances: Editions OCEP, 1981.

Villain, J. *La France moderne généalogique. Haute-Garonne et l'Ariège*, vol. 2. Montpellier: Firmin, Montane, and Sicardi, 1911.

Ville de Cadours. *Concours agricole cantonal, 9 et 10 séptembre 1967. Programme officiel*. [Cadours], 1967.

Warner, Charles K. *The Winegrowers of France and the Government Since 1875*. Westport, Conn.: Greenwood Press, 1975.

Weber, Eugen. "Comment la politique vint aux paysans: A Second Look at Peasant Politicization." *American Historical Review*, 87 (1982): 357–89.

————. *Peasants into Frenchmen: The Modernization of Rural France, 1870–1914*. Stanford, Calif.: Stanford University Press, 1976.

————. "The Second French Republic, Politics and the Peasant." *French Historical Studies* 11 (1980): 521–50.

Wolff, Philippe, ed. *Histoire de Toulouse* by Michel Labrousse, Philippe Wolff, Marcel Durliat, Bartolomé Bennassar, Bruno Tollon, Jacques Godechot. Toulouse: Privat, 1974.

Wright, Gordon. *Rural Revolution in France: The Peasantry in the Twentieth Century.* Stanford, Calif.: Stanford University Press, 1964.